SpringerBriefs in Molecular Science

For further volumes:
http://www.springer.com/series/8898

Yutaka Matsuo · Masayoshi Higuchi
Yuichi Negishi · Michito Yoshizawa
Takashi Uemura · Hikaru Takaya
Takafumi Ueno · Masayuki Takeuchi
Soichiro Yoshimoto
Editors

Metal–Molecular Assembly for Functional Materials

Springer

Editors

Yutaka Matsuo
Department of Chemistry
The University of Tokyo, Tokyo, Japan

Masayoshi Higuchi
Electronic Functional Materials Group
National Institute for Materials Science
Tsukuba, Japan

Yuichi Negishi
Faculty of Science
Department of Applied Chemistry
Tokyo University of Science
Tokyo, Japan

Michito Yoshizawa
Chemical Resources Laboratory
Tokyo Institute of Technology
Yokohama, Japan

Takashi Uemura
Department of Synthetic Chemistry and
  Biological Chemistry
Graduate School of Engineering
Kyoto University, Kyoto, Japan

Hikaru Takaya
Institute for Chemical Research
Kyoto University
Uji, Japan

Takafumi Ueno
Graduate School of Bioscience and
  Biotechnology
Tokyo Institute of Technology
Yokohama, Japan

Masayuki Takeuchi
Organic Materials Group
National Institute for Materials Science
Tsukuba, Japan

Soichiro Yoshimoto
Priority Organization for Innovation and
  Excellence
Kumamoto Institute for Photo-Electro
  Organics (Phoenics)
Kumamoto University, Kumamoto, Japan

ISSN 2191-5407   ISSN 2191-5415 (electronic)
ISBN 978-4-431-54369-5 ISBN 978-4-431-54370-1 (eBook)
DOI 10.1007/978-4-431-54370-1
Springer Tokyo Heidelberg New York Dordrecht London

Library of Congress Control Number: 2013943463

Printed on acid-free paper

Springer is part of Springer Science+Business Media (www.springer.com)

Preface

Coordination chemistry is now recognized as an essential area of research in modern nanoscience, because the molecular assembly techniques derived from coordination chemistry are applicable to the spontaneous build-up of nanostructures. In particular, recent progress in coordination chemistry has allowed for precise control over the structure and function of materials fabricated at the nanolevel. The dimensions of the structures assembled in this way can also be controlled by making changes to the "modules". For example, one-dimensional (1-D) linear polymer structures can be obtained by the complexation of ditopic organic ligands, such as bis(terpyridine), with transition metal ions that have an octahedral coordination space, such as Fe(II) or Ru(II). In contrast, 3-D framework structures such as porous coordination polymers can be formed by changing the organic ligands in these materials to bipyridine. Interestingly, this technique can also create discrete 3-D molecules, which can be used to extract specific compounds into their "nanospace" cavities. Regarding these cavities, larger 3-D cavities generally exist in proteins such as ferritin, which is itself composed of smaller proteins. Self-assembled structures of this particular type can also behave as cages for metal clusters in living systems. To prepare metal clusters with the required level of precision, it is particularly important to understand the electronic properties of the materials. The existence of the "magic number" concept in metal clusters was revealed by the application of the bottom-up self-assembly technique. Metal–metal or metal–organic interactions create unique 3-D structural assemblies with gel-like properties. Moreover, conventional conjugated polymer chains can be aligned in a 2-D manner using "clipping metal complexes". Self-assembled molecular packing with a 2-D structure can be obtained via the use of regularly ordered metal surfaces, and the resulting materials can be observed directly by scanning tunneling microscopy. In terms of their applications, self-assembled structures can be used to control the directional flow of electrons in electronic devices. Thus, fullerene layers have been used to good effect in organic photovoltaic devices, whereas a 1-D coordination polymer film on an indium tin oxide electrode has demonstrated electrochromic behavior.

This book focuses on modern coordination chemistry, with particular emphasis on porous coordination polymers, metalloproteins, metallopeptides, nanoclusters, nanocapsules, aligned polymers, and fullerenes. Furthermore, the book deals with the application of these materials to electronic devices and surface characterization. These wide-ranging topics have all been described in the context of dimensionality (1-, 2-, and 3-D), the design of new materials, synthesis, molecular assembly, function, and application. The 9 chapters making up this book have been authored by scientists who are at the cutting edge of research in this particular field and belong to the new frontier research group of "Metal and Assembly" in the Chemical Society of Japan.

Tokyo and Tsukuba, March 2013   Yutaka Matsuo
                                                                     Masayoshi Higuchi

Contents

Chapter 1
Molecular Assembly and Organization of Fullerenes for Photoelectric Conversion

Yutaka Matsuo

Abstract Functional fullerene derivatives bearing five feathers of long alkyl chains give rise to a one-dimensional columnar alignment in the liquid crystalline state. Fullerene derivatives with five legs composed of anchoring carboxylic acid groups form self-assembled monolayers on indium-tin-oxide electrodes to afford photocurrent generating cells for photosensitive switching devices. Bis(silylmethyl)fullerene forms a three-dimensional honeycomb-like structure following thermal crystallization in thin films, which can be employed in the construction of high performance organic solar cells.

Keywords Fullerene • Liquid crystals • Self-assembled monolayers • Photovoltaics • Organic solar cells

1.1 Introduction

The construction of well-ordered structures composed of fullerene derivatives not only in the bulk, but also on the surface is essential for the creation of a variety of different functional devices. Fullerene, however, is spherical in shape, and tends to form aggregates, which makes it difficult to construct well-ordered structures. This chapter will focus on the assembly of fullerene derivatives to form 1-D, 2-D, and 3-D organization structures in both the solid state and on the surface. Particular emphasis will be given to descriptions of the liquid crystals and self-assembled monolayers (SAMs) of fullerene derivatives with five feathers or legs attached to the fullerene core, as well as the columnar fullerene alignment triggered by thermal crystallization. During the development of

Y. Matsuo (✉)
Department of Chemistry, School of Science, The University of Tokyo, Tokyo, Japan
e-mail: matsuo@chem.s.u-tokyo.ac.jp

Y. Matsuo et al. (eds.), *Metal–Molecular Assembly for Functional Materials*,
SpringerBriefs in Molecular Science, DOI: 10.1007/978-4-431-54370-1_1,
© The Author(s) 2013

high performance organic electronic devices, it is essential that adequate consideration is given to the arrangement of the fullerene derivatives to allow for the construction of well-defined structures within these devices. The application of these materials to organic solar cells using self-assembling fullerene derivatives has also been described in this chapter with a view to providing an understanding of the rationale behind the design of device structures for high levels of efficiency.

1.2 Liquid Crystalline Fullerene Derivatives

Fullerene-containing liquid crystals are promising electro/photo active soft materials. The development of techniques for the elaboration of liquid crystalline fullerene derivatives, however, represents a significant synthetic challenge, which is limited by the size and shape of the fullerene unit. Badminton-shuttlecock-shaped molecules (Fig. 1.1, compound **1**), which have five organic "feathers" attached to the [60]fullerene unit, undergo an assembly process to form one-dimensional columnar structures via head-to-tail stacking (Fig. 1.1) [1]. Molecules bearing long alkyl side chains form thermotropic hexagonal columnar liquid crystals when the alkyl chains melt at temperature in the range of 0–140 °C. Such supramolecular assembly processes occur as a result of three different driving forces, including (1) microphase segregation between the aromatic fullerene cores and the aliphatic alkyl chains; (2) the recognition of molecular shape in which ball-shaped fullerene cores are accommodated by the cone-shaped cavities formed from the five organic addends; and (3) π–π stacking interactions between the fullerene cores and the phenyl groups. The expansion of the cone-shaped cavity into a cup-shaped cavity affords an enhanced level of stability to the columnar stacking, and leads to the formation of liquid crystals over a much wider range of temperatures (0–180 °C, compound **2**, Fig. 1.2) [2]. Redox-active donor–acceptor-type fullerene liquid crystals can be obtained via the installation of

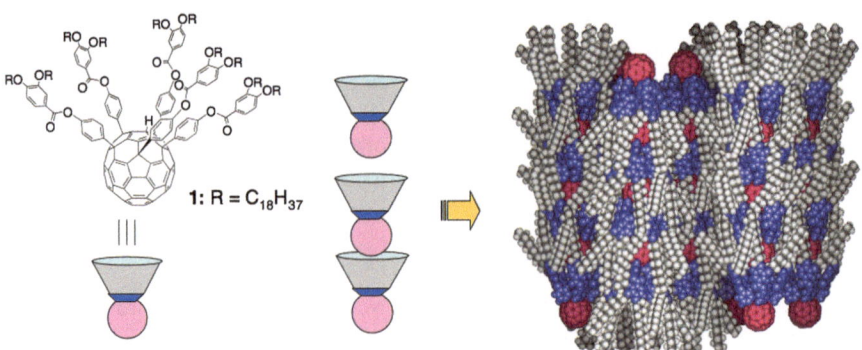

1: R = $C_{18}H_{37}$

Fig. 1.1 Columnar liquid crystalline assembly of shuttlecock-shaped fullerenes

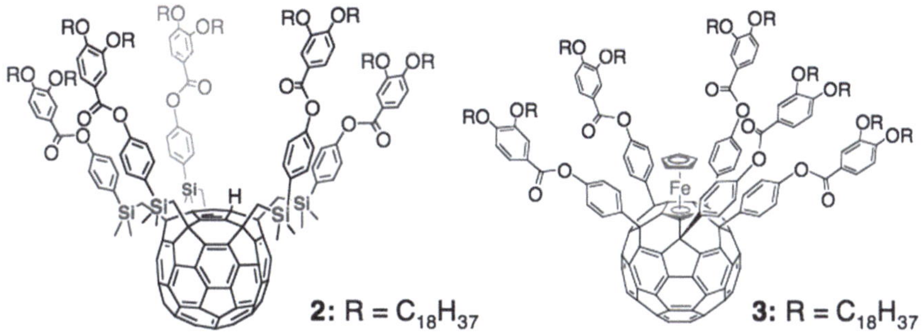

2: R = C_{18}H_{37}

3: R = C_{18}H_{37}

Fig. 1.2 Molecular structure of liquid crystalline fullerene molecules

a ferrocene moiety in the cavity (compound **3**, Fig. 1.2) [3]. These compounds can undergo a reversible three-electron reduction process at their fullerene moiety, as well as a one-electron reduction process at their ferrocene moiety.

Changes to the molecular shape can afford different supramolecular organization properties. Smectic liquid crystals with layer-by-layer structures (Fig. 1.3) can be obtained by effectively breaking the cone-shaped cavity via the replacement of the five feathers with five rod-shaped structures, which reduces the number of phenyl groups on the five organic addends from ten to five (compound **4**) [4]. As a result of these changes, the structure formed from the five rod-shaped addends

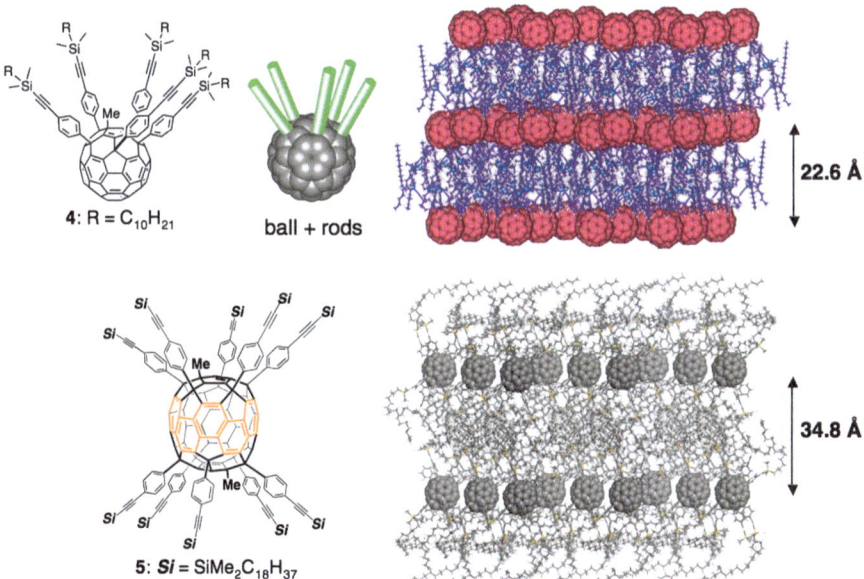

4: R = C_{10}H_{21}   ball + rods   22.6 Å

5: Si = SiMe_2C_{18}H_{37}   34.8 Å

Fig. 1.3 Layer-by-layer liquid crystalline assembly of fullerene derivatives

cannot accommodate the fullerene moiety of another molecule. A significant body of positive design work has been conducted using deca-aryl fullerenes. Fullerene deca-adducts (compound **5**) bearing ten organic addends over the top and bottom of fullerene also give layered structures in liquid crystals (Fig. 1.3) [5]. The smectic liquid crystalline characteristics of this material were determined by X-ray diffraction analysis, which revealed the interlayer distance to be 3.5 nm. Analysis by differential scanning calorimetry revealed that the liquid crystalline state existed over a temperature range of 10–250 °C. Compound **5** was found to be luminescent and emit yellow fluorescence upon excitation because it has a conjugated cyclic π-electron system at its center. The quantum yields for the luminescence of **5** in the solution and solid states were 21 and 18 %, respectively. Interestingly, polarized emission has been observed in liquid crystalline thin films following the application of a mechanical rubbing treatment.

Cubic-like liquid crystals have been obtained using a polarized fullerene molecule bearing long alkyl chains (compound **6**) [6]. This polar fullerene compound was constructed with a ferrocene-fullerene fused structure, where the ferrocene and fullerene moieties were responsible for positive and negative aspects of the compound, respectively. Four such dipolar molecules can form octupolar aggregates (Fig. 1.4), which have melting alkyl chains in their shell region that enable the aggregate to rotate. The liquid crystalline properties of these materials have been well characterized across a wide range of temperatures from the freezing of the alkyl chains at 55 °C to their isotropization through the breaking of the aggregates at 230 °C.

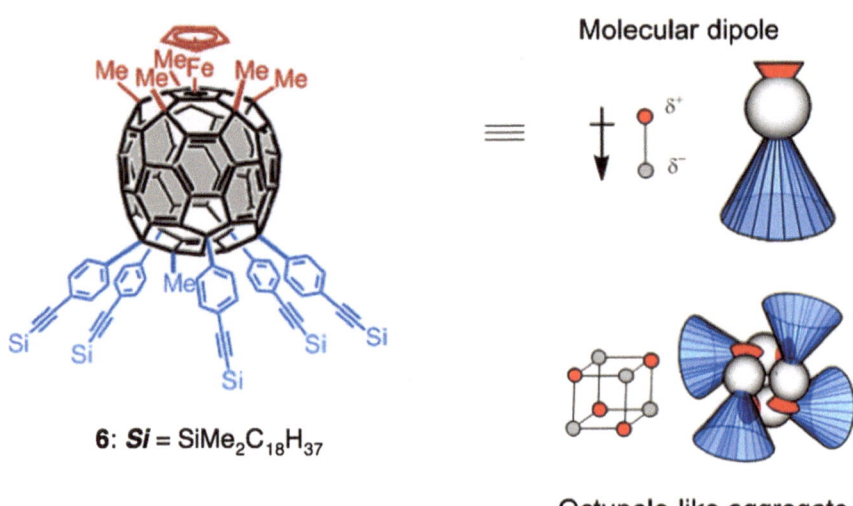

Fig. 1.4 Cubic-like liquid crystalline assembly though the octupolar aggregation of polar fullerene derivatives

1.3 Self-Assembled Monolayers of Fullerene Derivatives

Fullerene is one of several intriguing materials that can be used for the construction of photocurrent generating photoelectrochemical cells, because it shows high electron affinity and a long-lived excited state upon photoabsorption. The penta-pod fullerene derivatives **7–9** (Fig. 1.5), which are so-named because they are effectively bearing five "legs", have been synthesized to immobilize the molecules in a 2-D manner on indium-tin oxide (ITO) electrodes (Fig. 1.5) [7]. The pentapod fullerenes resemble a lunar landing module in shape. The preparation of SAMs from the pentapod molecules is typically performed via the immersion of ITO substrates into a THF solution of the pentapod molecules. The direction of the photocurrent can be switched by the components (e.g., methylated compounds or iron complexes) and the molecular orientation (i.e., standing upright or lying down). The photocurrent generation mechanism involves electron transfer between the

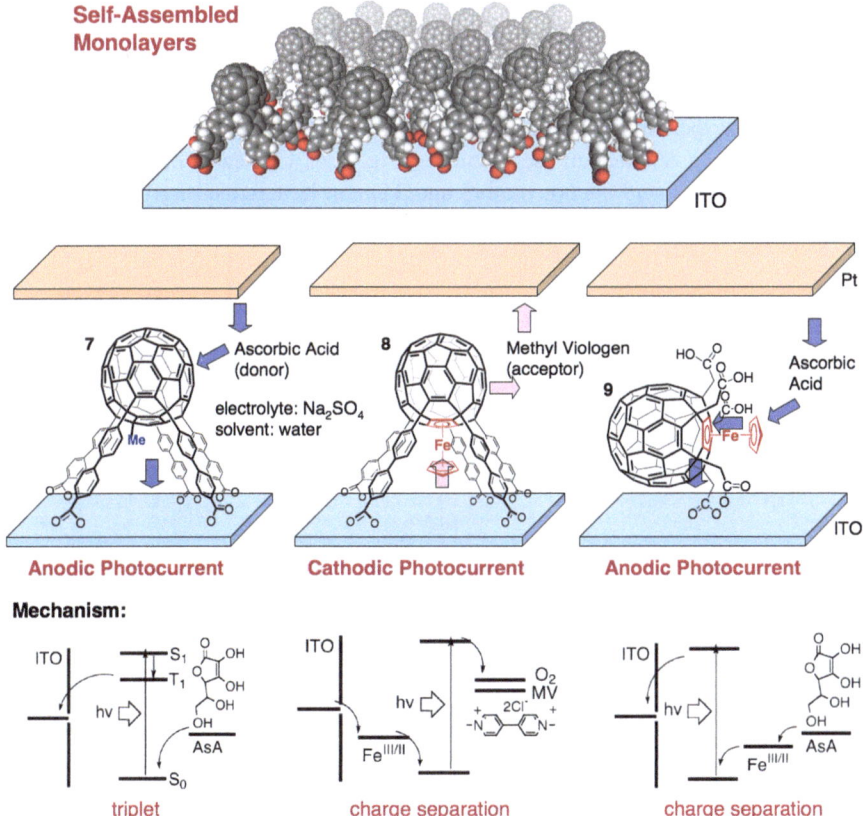

Fig. 1.5 Photocurrent-generating SAMs of pentapod fullerene derivatives

fullerene molecules, electrodes, and the coexisting donor/acceptor moieties. The methylated compound **7** can be excited to the triplet state to receive an electron from an ascorbic acid donor to form a fullerene radical anion, which can give an electron to the ITO electrode to generate an anodic photocurrent from the platinum counter electrode to the ITO working electrode. In the case of the iron complex **8**, an electron transfer can occur from the ferrocene moiety to the fullerene moiety under light irradiation to generate a ferrocenium cation and a fullerene radical anion. The ferrocenium cation can receive an electron from the ITO electrode, whereas the fullerene radical anion can give an electron to the methyl viologen acceptor to generate a cathodic photocurrent from the ITO electrode to the platinum electrode. As for compound **9**, the ferrocene part points outwards and away from the ITO surface. The ferrocenium cation can receive an electron from the donor, whereas the fullerene radical anion can give an electron to the ITO electrode to generate an anodic photocurrent.

Mixed component SAMs consisting of a C_{60}-non metallic compound (compound **7**) and a C_{70}-iron complex (compound **10**) have demonstrated a switching device generating anodic and cathodic photocurrents under shorter and longer wavelengths of light, respectively (Fig. 1.6) [8]. This system has two optical inputs (shorter and longer wavelengths of the light) and two electronic outputs (anodic and cathodic photocurrents), and thus represents an optical/electronic interface that is capable of transforming optical information into electronic information.

As well as pentapod fullerenes, several other self-assembling fullerene molecules have been reported in the literature (Fig. 1.7). Conical fullerene derivatives with a phosphonic acid anchor (compound **11**) can be fixed strongly onto the ITO

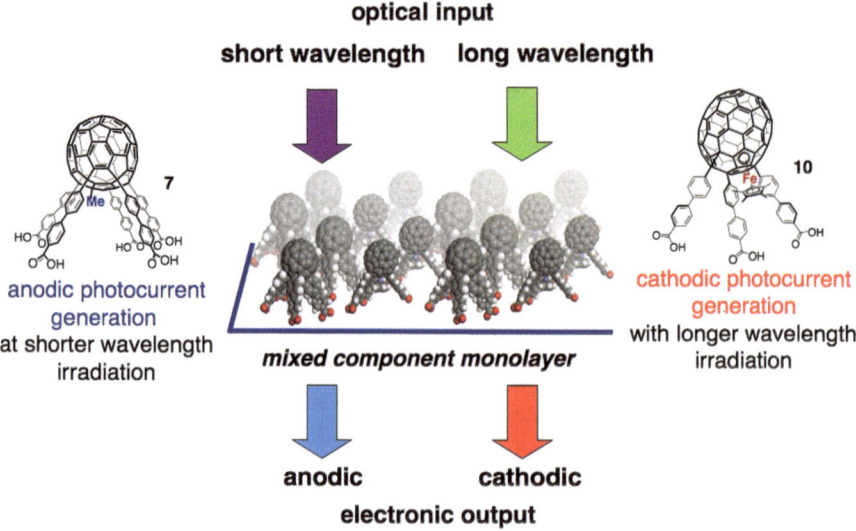

Fig. 1.6 Bidirectional photocurrent generation systems switched by the wavelength of the light

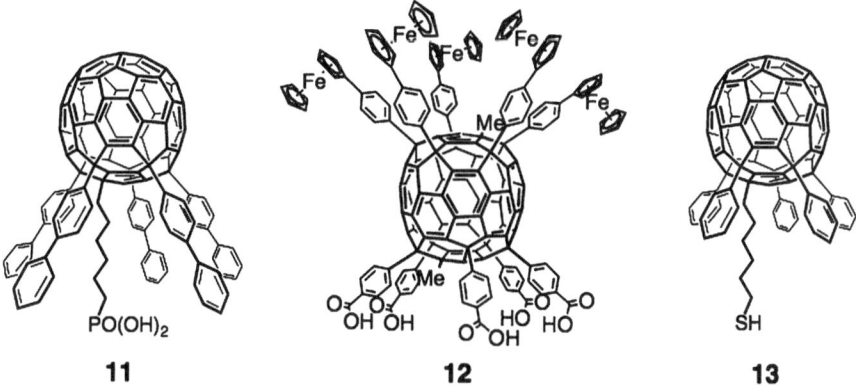

Fig. 1.7 Molecular structure of two-dimensionally self-assembling fullerene molecules

surface [9]. Pentapod deca-aryl fullerenes bearing functionalities on the top part of their fullerene moiety (compound **12**) can form bilayers or multilayers on the surface of the ITO electrode to generate a larger photocurrents than those observed in the corresponding monolayer systems [10]. Fullerene-thiol molecules **13** form SAMs on gold substrates, where relatively longer alkyl linkers connecting the fullerene and thiol parts, compared with the shorter penta-aryl spacers, are necessary to obtain monolayers of the fullerene-thiols [11]. Umbrella-shaped fullerene molecules have been used to modify the surfaces of electrodes and change the work function of the electrodes [12]. Supramolecular photocurrent-generating systems have been constructed using magnesium porphyrin–fullerene linked molecules, which are fixed to the SAM via an imidazole carboxylic acid on the ITO [13].

1.4 Alignment of Fullerene Derivatives in the Solid State for Photovoltaic Applications

The 3-D assembly of crystals has been seen in the thermal crystallization of bis(dimethylphenylsilylmethyl)fullerene (SIMEF) (Fig. 1.8) [14]. SIMEF has crystallization and melting temperatures of 150 and 225 °C, respectively. Amorphous thin films of SIMEF on substrates are typically heated to 150 °C to obtain crystalline thin films, which generally possess a higher level of electron mobility than the corresponding amorphous materials. In fact, SIMEF shows a respectable electron mobility value of 8×10^{-3} cm^2/V s under the space-charge limited current theory, which is higher than that of the commonly used fullerene electron-acceptor, [6,6]-phenyl-C$_{61}$-butyric acid methyl ester (6×10^{-3} cm^2/V s). In the crystalline state, the fullerene cores of the SIMEF molecules align in a 1-D manner and form a 3-D distorted honeycomb-like structure. This structure is formed by phase separation between the rigid aromatic fullerene cores and the flexible silylmethyl side

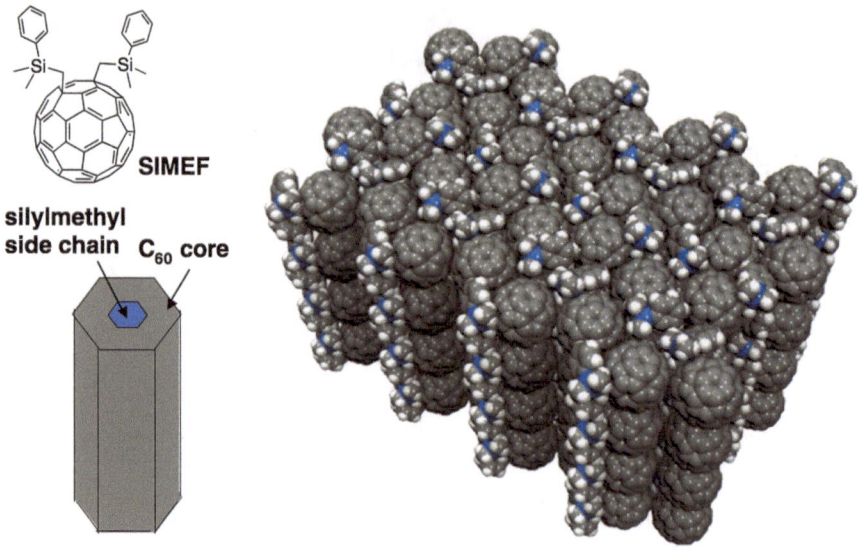

Fig. 1.8 Molecular structure of SIMEF and its crystal packing structure

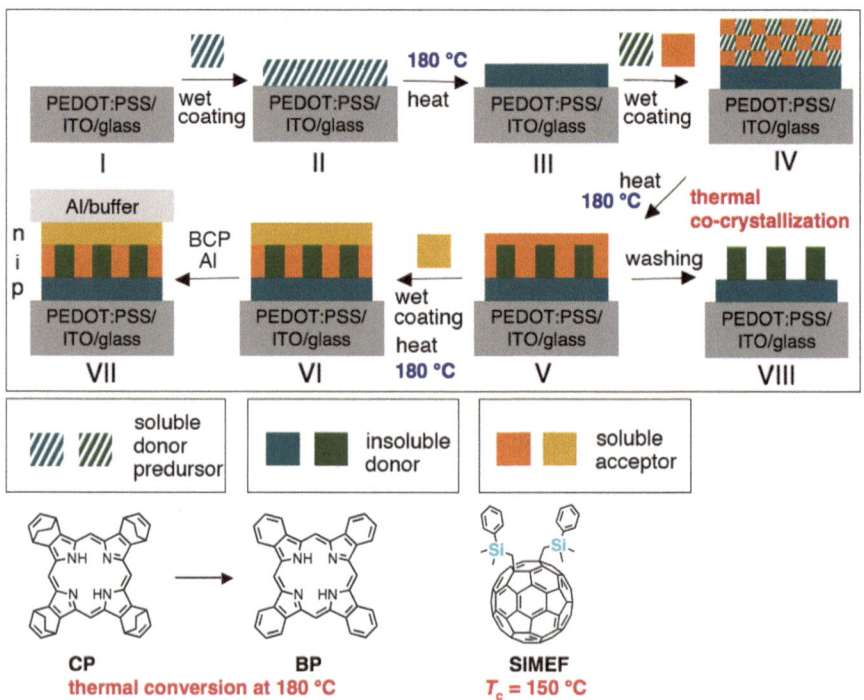

Fig. 1.9 Process for solution-processed BP:SIMEF organic solar cells

chain units. The thermal crystallization and alignment processes of these fullerene cores are used to construct interpenetrating electron-donor/acceptor interface structures in organic photovoltaic devices.

SIMEF has been employed in organic thin-film solar cells as a good electron acceptor [15–17]. Solution-processed small-molecule-based organic solar cells can be constructed using a soluble tetrabenzoporphyrin precursor (CP) as an electron donor and SIMEF as an electron acceptor. A solution composed of a mixture of CP and SIMEF is deposited onto the tetrabenzoporphyrin (BP) layer (Fig. 1.9, device structure III) by spin-coating to give the blended thin film of CP and SIMEF (structure IV). The substrate is then heated to 180 °C to affect the thermal co-crystallization process involving the thermal conversion of CP into BP and the thermal crystallization of SIMEF. During this process, both materials crystallize simultaneously in the thin film to give a phase-separated structure containing fine columnar crystals of BP (25 nm diameter, 65 nm height), with SIMEF distributed among the columnar BP crystals. This represents an ideal structure for both charge separation and carrier (electron and hole) transport. This structure was characterized by scanning electron microscopy using the substrate following the washing out of SIMEF (structure VIII). Organic solar cells (structure VII) showed a 5.2 % power conversion efficiency with a short-circuit current density of 10.5 mA/cm^2, open-circuit voltage of 0.75 V, and fill factor of 0.65.

Acknowledgments The author would like to extend their sincerest thanks to Professor Eiichi Nakamura (The University of Tokyo), who was an important collaborator on the work mentioned above.

References

1. Sawamura M, Kawai K, Matsuo Y, Kanie K, Kato T, Nakamura E (2002) Stacking of conical molecules with a fullerene apex into polar columns in crystals and liquid crystals. Nature 419:702–705. doi:10.1038/nature01110
2. Matsuo Y, Muramatsu A, Hamasaki R, Mizoshita N, Kato T, Nakamura E (2004) Stacking of molecules possessing a fullerene apex and a cup-shaped cavity connected by silicon-connection. J Am Chem Soc 126:432–433. doi:10.1021/ja038816y
3. Matsuo Y, Muramatsu A, Kamikawa Y, Kato T, Nakamura E (2006) Synthesis, structural, electrochemical and stacking properties of conical molecules possessing buckyferrocene on apex. J Am Chem Soc 128:9586–9587. doi:10.1021/ja062757h
4. Zhong YW, Matsuo Y, Nakamura E (2007) Lamellar assembly of conical molecules possessing a fullerene apex in crystals and liquid crystals. J Am Chem Soc 129:3052–3053. doi:10.1021/ja068780k
5. Li CZ, Matsuo Y, Nakamura E (2009) Luminescent bow-tie-shaped decaaryl[60]fullerene mesogens. J Am Chem Soc 131:17058–17059. doi:10.1021/ja907908m
6. Li CZ, Matsuo Y, Nakamura E (2010) Octupole-like supramolecular aggregates of conical iron fullerene complexes into a three-dimensional liquid crystalline lattice. J Am Chem Soc 132:15514–15515. doi:10.1021/ja1073933
7. Matsuo Y, Kanaizuka K. Matsuo K, Zhong YW, Nakae T, Nakamura E (2008) Photocurrent-generating properties of organometallic fullerene molecules on an electrode. J Am Chem Soc 130:5016–5017. doi:10.1021/ja800481d

8. Matsuo Y, Ichiki T, Nakamura E (2011) Molecular photoelectric switch using a mixed sam of organic [60]fullerene and [70]fullerene doped with a single iron atom. J Am Chem Soc 133:9932–9937. doi:10.1021/ja203224d

9. Sakamoto A, Matsuo Y, Matsuo K, Nakamura E (2009) Efficient bidirectional photocurrent generation by self-assembled monolayer of penta(aryl)[60]fullerene phosphonic acid. Chem Asian J 4:1208–1212. doi:10.1002/asia.200900155

10. Matsuo Y, Ichiki T, Radhakrishnan SS, Guldi DM, Nakamura E (2010) Loading pentapod deca(organo)[60]fullerenes with electron donors: from photophysics to photoelectrochemical bilayers. J Am Chem Soc 132:6342–6348. doi:10.1021/ja909970h

11. Matsuo Y, Lacher S, Sakamoto A, Matsuo K, Nakamura E (2010) Conical pentaaryl[60] fullerene thiols: self-assembled monolayers on gold and photocurrent generating property. J Phys Chem C 114:17741–17752. doi:10.1021/jp1059402

12. Lacher S, Matsuo Y, Nakamura E (2011) Molecular and supramolecular control of the work function of an inorganic electrode with self-assembled umbrella-shaped fullerene derivatives. J Am Chem Soc 133:16997–17004. doi:10.1021/ja206767

13. Ichiki T, Matsuo Y, Nakamura E (2013) Photostability of a dyad of magnesium porphyrin and fullerene and its application to photocurrent conversion. Chem Commun 49:279–281. doi:10.1039/C2CC36988E

14. Matsuo Y, Iwashita A, Abe Y, Li CZ, Matsuo K, Hashiguchi M, Nakamura E (2008) Regioselective synthesis of 1,4-di(organo)[60]fullerenes through DMF-assisted mono-addition of silylmethyl Grignard reagents and subsequent alkylation reaction. J Am Chem Soc 130:15429–15436. doi:10.1021/ja8041299

15. Matsuo Y, Sato Y, Niinomi T, Soga I, Tanaka H, Nakamura E (2009) Columnar structure in bulk heterojunction in solution-processable three-layered p-i-n organic photovoltaic devices using tetrabenzoporphyrin precursor and silylmethyl[60]fullerene. J Am Chem Soc 131:16048–16050. doi:10.1021/ja9048702

16. Matsuo Y, Hatano J, Kuwabara T, Takahashi K (2012) Fullerene acceptor for improving open-circuit voltage in inverted organic photovoltaic devices without accompanying decrease in short-circuit current density. Appl Phys Lett 100:063303. doi:10.1063/1.3683469

17. Tanaka H, Abe Y, Matsuo Y, Kawai J, Soga I, Sato Y, Nakamura E (2012) An amorphous mesophase generated by thermal annealing for high-performance organic photovoltaic devices. Adv Mater 24:3521–3525. doi:10.1002/adma.201200490

Chapter 2
Organic–Metallic Hybrid Polymers for Electrochromic Display Devices

Masayoshi Higuchi

Abstract Organic–metallic hybrid polymers can be synthesized via the simple complexation of transition metal ions such as Fe(II) and Ru(II) ions with organic ligands bearing two coordination sites. The resulting hybrid polymers have specific colors based on their metal-to-ligand charge transfer (MLCT) absorption. Polymer films can be obtained using the standard methods of preparation, such as spin coating. Interestingly, the appearance/disappearance of the color from the polymer film can be controlled by the electrochemical reduction/oxidation (redox) of the metal ions in the polymer film, because the bandgap of the MLCT absorption changes during the redox process. A variety of different organic–metallic hybrid polymers can be obtained using different combinations of organic ligands and metal ions. Electrochromic solid-state devices have been successfully fabricated using the organic–metallic hybrid polymer films and gel electrolytes. These electrochromic devices have the potential to be applied for the development of next generation displays, including color electronic papers and smart windows.

Keywords Organic–metallic hybrid polymer • Metallosupramolecular polymer • Complexation • Electrochromic property • Display device

2.1 Introduction

Recent developments in the field of telecommunications, including the internet and cellular phones, have resulted in our daily lives becoming more comfortable than ever before. As a consequence, however, the amount of carbon dioxide released from electronic devices such as computers and visual display units

M. Higuchi (✉)
Electronic Functional Materials Group, National Institute for Materials Science, Tsukuba, Japan
e-mail: HIGUCHI.Masayoshi@nims.go.jp

Y. Matsuo et al. (eds.), *Metal–Molecular Assembly for Functional Materials*, SpringerBriefs in Molecular Science, DOI: 10.1007/978-4-431-54370-1_2, © The Author(s) 2013

is constantly increasing. Given this scenario, it is difficult to disregard problems pertaining to energy consumption and global warming. Liquid-crystal, plasma, and organic electroluminescent (EL) displays currently require a source of electricity to continue to display. In addition, although it was envisaged that the increasing popularity of personal computers would result in a reduction in paper usage, people have continued to print documents and pictures from their computers, perhaps because many people still feel more comfortable reading text that has been printed on paper than reading text off a computer screen. To overcome these problems pertaining to energy consumption, the environment, and the use of resources, electronic papers [1–3] have recently been introduced as a new form of display. Electronic papers resemble ordinary paper, in that they have good flexibility, are thin and lightweight, and display characters and images, albeit electronically. These displays reflect light in a manner identical to that of ordinary paper and therefore do not require power to display text and images, that is, the text and images do not disappear even after the power has been turned off. These displays are also portable and only require a power supply when the display information is being rewritten. Therefore, in an era in which every industry and utility is driven by the keywords "energy conservation" and "power saving", electronic paper could prove to be a significant display device. It is envisaged that electronic papers could ultimately replace normal printed papers such as newspapers and posters.

A variety of different methods have been proposed as candidates to develop the display mechanism of electronic paper, including the microcapsule method. It is noteworthy, however, that most of these methods are only suitable for the potential fabrication of monochromatic (black and white) displays. Electrochromic displays [4, 5], which involve the use of electrochromic materials that change color as a consequence of electrochemical oxidation and reduction processes, are known to be suitable for the construction of electronic papers with color displays. Unfortunately, however, research on these displays has been limited compared with research on other methods. There are several reasons why research into color displays has been limited, including (1) there are several drawbacks associated with the properties of this material, such as low durability, slow response, and difficulties associated with its use during the fabrication of solid-state devices; and (2) the efficiency of the method is heavily dependent on the superiority or inferiority of the electrochromic material. In other words, if an electrochromic material is developed that is suitable for use in electronic paper, it is possible that this method could be a breakthrough in the area of color electronic paper.

Molybdenum oxide, Prussian blue, and phthalocyanine metal complexes are well known as inorganic electrochromic materials, whereas viologen derivatives and conductive polymers are well known examples of organic electrochromic materials. Interestingly, however, only a limited number of applications of electrochromic materials have been identified. The limited application of these materials has been attributed to the fact that inorganic electrochromic materials offer only a limited range of color variation, and that their use involves complicated processing techniques. In contrast, organic electrochromic materials exhibit a variety of different colors, but they are poorly durable. For example, organic conductive polymers such as polyaniline have been comprehensively studied as electrochromic materials over the last two decades, and several applications of these materials have been proposed, with the majority of these applications being largely impractical. These polymers are considered to be impractical primarily because they possess low levels of stability for the color change. When the polymers are injected with charges (or electrons), a color change is observed. The electrochromic properties are caused by changes in the electronic state and structure of the polymers. As a result, even if the original polymer is stable, the structure of the polymer following oxidization (reduction) can be changed to the extent that the material becomes unstable to light, heat, or air, and subsequently deteriorates as the electrochemical redox switching is operated repeatedly. To identify practical applications for these materials, electrochromic materials must be developed that are both stable and reliable. In this respect, organic electrochromic materials are currently at a major disadvantage.

2.2 Organic–Metallic Hybrid Polymers

To develop new electrochromic material with high durability, which would be particularly advantageous in inorganic electrochromic materials, and abundant color variation, which would be particularly advantageous in organic electrochromic materials, our own research has focused specifically on hybrid polymers bearing both organic and metallic parts, that is, organic–metallic hybrid polymers [6-15]. The organic–metallic hybrid polymers of interest were synthesized via the complexation of metal ions with organic ligands bearing two coordination sites, such as bis(terpyridine) type ligands. The organic ligands and metal ions of the resulting hybrid polymers were found to be alternately connected throughout the polymeric structure (Fig. 2.1). Although polymers of this particular type have already been reported as metallosupramolecular polymers by Hofmeier and Schubert [16], the electrochromic properties of these materials have not yet been thoroughly investigated.

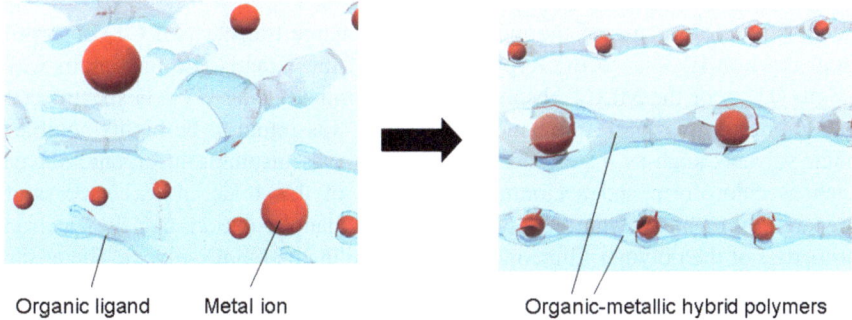

Organic ligand   Metal ion   Organic-metallic hybrid polymers

Fig. 2.1 Schematic presentation of organic–metallic hybrid polymer formation by complexation of the metal ions with organic ligands

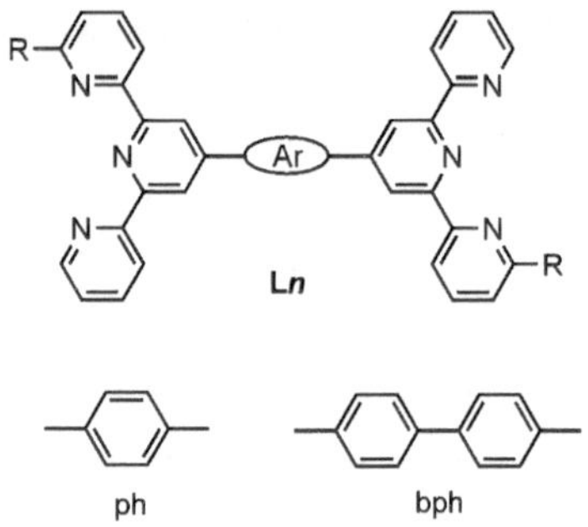

Fig. 2.2 Bis(terpyridine) derivatives (**L1–5**) used in the synthesis of organic–metallic hybrid polymers

L*n*	R	Ar
L1	H	ph
L2	H	bph
L3	OMe	ph
L4	OMe	bph
L5	Br	ph

One of the advantages of using this technique to synthesize the polymers is the relative ease with which the polymer chains can be formed through the complexation between the metal ions and the organic ligands. In a typical procedure for the synthesis of the polymers, equimolar amounts of 1,4-bis(2,2′:6′,2″-terpyridin-4-yl) benzene (**L1**) and iron(II) acetate are refluxed in argon-saturated acetic acid (ca. 10 mL of solvent per 1 mg of **L1**) for 24 h. As the complexation process progresses, the color of the solution becomes purple. Upon completion of the reaction, the solution is cooled to ambient temperature and filtered to remove any insoluble residues. The filtrates are then placed in a Petri dish, and the solvent allowed to slowly evaporate until the material is dry. The resulting brittle film can then be collected and dried further *in vacuo* overnight to obtain **FeL1** (yield > 90 %). The other hybrid polymers (**FeL***n* (*n* = 1–5)) were also obtained in the same way by the complexation of **L1–5** (Fig. 2.2) with iron(II) acetate in acetic acid. A variety of different polymers can be prepared by changing the organic ligand or metal species. The use of ethylene glycol as the solvent and Ru(DMSO)$_4$Cl$_2$ as the Ru(II) precursor, allowed for a high yield (>95 %) of high-purity **RuL***n* to be obtained.

2.3 Electrochromic Properties

FeL1 is purple because of the metal-to-ligand charge transfer (MLCT) absorption from the iron(II) ions to the terpyridyl units of the ligand. The maximum wavelength (λ_{max}) of the MLCT absorption, as determined using UV–vis spectroscopic measurements, was 580 nm in methanol. **FeL1** was found to be highly soluble in polar solvents such as water and methanol, but was insoluble in organic solvents such as chloroform and acetonitrile. The film of this material was successfully prepared by means of spin coating from the methanol solution, and the electronic properties of the polymer film could be measured in an organic solvent. The electrochemical properties were investigated by cyclic voltammetry using a glassy carbon electrode that had been covered with a thin film of **FeL1** as the working electrode. The **FeL1** film showed a large current response when it was oxidized in an acetonitrile solution. The reduction/oxidation (redox) activity of the film was completely reversible ($\Delta E = 75$ mV at 0.1 V/s scan rate). The half-wave potential ($E_{1/2}$) of the redox reaction was 0.77 V versus Ag/Ag^+. The electrochemical redox in the **FeL1** film was based on the redox reaction between iron(II) and iron(III) [17–20].

Interestingly, when the polymer film was cast onto an indium tin oxide (ITO) electrode using methanol, the material exhibited electrochromic properties in acetonitrile, in that the blue-colored polymer film became almost colorless when a voltage of 1.0 V was applied to the electrode (Fig. 2.3). Furthermore, the film regained its original blue color when it was subjected to a reduction voltage of 0 V. UV–vis analysis of the polymer film during the electrochemical oxidation revealed spectral change, which were monitored by means of in situ electrospectral measurements, where the UV–vis spectra of the polymer film cast onto an ITO transparent electrode were measured whilst simultaneously applying the voltage to the electrode in a quartz cuvette (1 mm-path length) with three necks. The disappearance of the MLCT absorption around 580 nm during the electrochemical oxidation process was investigated using this technique, which confirmed that the MLCT absorption disappeared at an oxidative potential of 1.0 V. The applied potential provided strong evidence that the color change in the polymer film occurred according to the electrochemical redox reaction of the iron ions. The MLCT absorption occurred in the presence of Fe(II) ions but disappeared in the case of **FeL1** which contained Fe(III) ions.

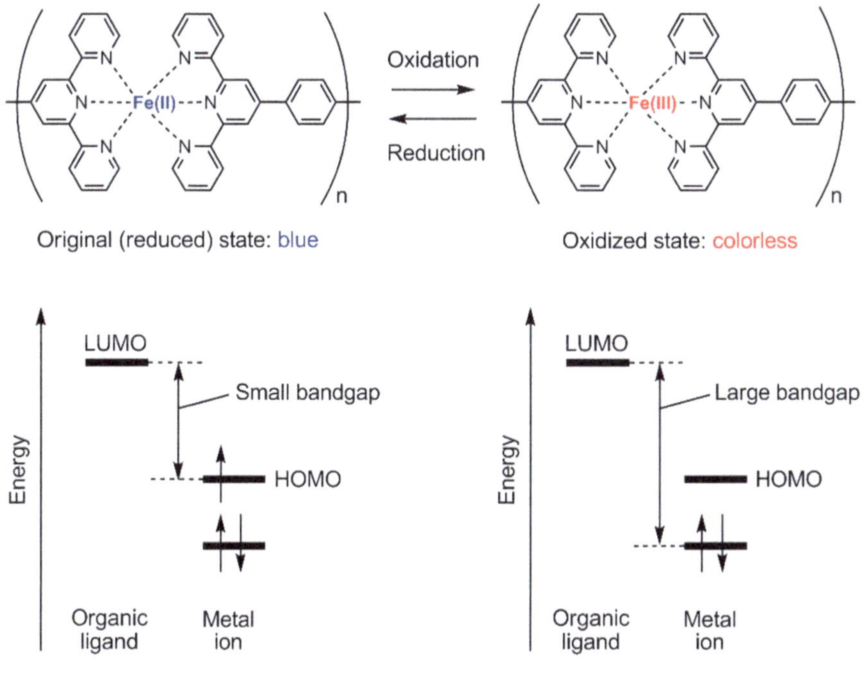

Fig. 2.3 Electrochromic behavior in a **FeL1** film and the mechanism

The colors of organic–metallic hybrid polymers are dependent on the potential gap between the lowest unoccupied molecular orbital (LUMO) of the organic ligand and the highest occupied molecular orbital (HOMO) of the metal ions. The MLCT absorption shifts toward shorter wavelengths when the potential gap increases and shifts towards longer wavelengths when the gap decreases. The colors of the hybrid polymers could be successfully adjusted by changing the metal species and/or by modifying the organic ligands with electron releasing groups (e.g., methoxy groups) or electron withdrawing groups (e.g., bromo groups). In addition, the use of different spacer units of different lengths in the bis(terpyridine) ligands also influenced the color of the polymers. As the result, the **FeL***n* polymers could be purple, blue, or green in color, whereas the cobalt(II) polymers (**CoL***n*), which were synthesized in a similar manner, were yellow or orange in color. The **RuL***n* polymer was red.

One of the main features of electrochromic hybrid polymers is that their transition from their colored form to their bleached form can be controlled by the electrochemical redox reactions of the metal ions in the polymers, because the color of the polymers appears as the complementary color of the absorption based on the charge transfer from the metal ions to the organic ligands. The electrochromic behavior in the polymer film can be explained using an energy diagram, as shown in Fig. 2.3. Since the HOMO potential of Fe(III) is lower than that of Fe(II), the potential gap of the MLCT absorption is increased by the electrochemical oxidation of the Fe(II) ions. As a result, the absorption moves from the visible light region to the invisible (ultraviolet) region and the polymer film becomes colorless.

The electrochromic properties of organic electrochromic materials generally deteriorate rapidly because the color changes in these materials occurs as a result of structural changes based on the electrochemical redox mechanism discussed above. The long-term stability of these materials is therefore low. Repetition stability can be significantly improved when organic–metallic hybrid polymers are used because the structures of the organic ligands do not change during the redox reactions of the metal ions. The results of an experiment involving the electrochromic cycling behavior of an **FeL1** film cast on ITO glass revealed that the response time did not change, even when the experiment was repeated 4,000 times. These results therefore indicate that the **FeL1** is highly durable towards repetitive electrochromic changes and that the major drawbacks associated with electrochromic organic materials that were outlined above have been overcome.

Given that organic–metallic hybrid polymers are formed through a self-assembly process, it is possible to not only select a variety of different combinations of organic ligands and metal species, but also to introduce several different types of organic ligands or metal species into a single polymer chain. For instance, an organic–metallic hybrid polymer containing both iron(II) and cobalt(II) ions (**Co/FeL1**) was constructed that changed color from red to blue to transparent/colorless at potentials of 0, 0.6, and 1.0 V, respectively. Multicolor electrochromic changes of this type occur as a result of differences in the redox potentials of the iron and cobalt ions (0.77 and 0.10 V vs Ag/Ag$^+$, respectively). The absorptions processes in which the iron(II) and cobalt(II) ions participated disappeared at potentials higher than these redox potentials. The initial red color observed was caused by a combination of two absorptions in which both the iron(II) and cobalt(II) ions participated. The color of the polymer changed to blue at a potential of 0.6 V. At this potential, the cobalt(II) ions were electrochemically oxidized to cobalt(III) ions, and the absorption in which the cobalt(II) ions participated therefore disappeared. Finally, the polymer became colorless at a potential of 1.0 V because the MLCT absorption involving iron(II) also disappeared as a result of the oxidation of iron(II) to iron(III). Such multicolor electrochromic polymers will assist in the simplification of devices such as multicolor electronic paper and in the reduction of the thickness of such devices.

2.4 Solid-State Display Devices

To date, although the electrochromic properties of polymer films have only been demonstrated in solution, solid-state devices that use electrochromic films could find practical applications in electronic devices and displays. Therefore, a solid-state electrochromic device was developed using a mixture of poly(methylmethacrylate) (PMMA), propylene carbonate (PC), and LiClO$_4$ as the polymer gel electrolyte (Fig. 2.4). For the fabrication of this device, 20 μL of a methanol solution (1.0 mg **FeL1**/0.5 mL MeOH) was cast or spin-coated (500 rpm for 60 s) onto an ITO glass electrode (20 Ω/sq). A mixture of PMMA (7.0 g), PC (20 mL), and LiClO$_4$ (3.0 g) was prepared as the polymer gel electrolyte. The gel electrolyte layer was located on the **FeL1** film, and another ITO glass was placed onto the electrolyte layer. When a potential of 3 or −3 V was applied between the two ITO electrodes, the **FeL1** film in the solid-state electrochromic device changed from colorless to blue, with the transition being particularly rapid (within one second) or relatively slow (a few seconds), respectively. The difference in the rate of the transition was attributed to the Fe(II) state being more stable than the Fe(III) state in the organic–metallic hybrid polymer. We also succeeded in fabricating 5- and 10-inch solid-state devices equipped with two polymer films, which could exhibit two

Fig. 2.4 A device structure of the electrochromic solid-state display using the organic–metallic hybrid polymer film

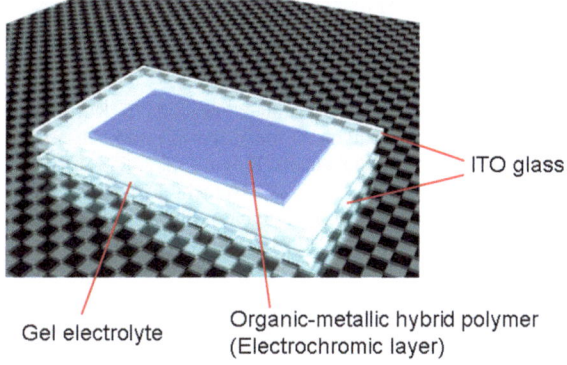

ITO glass

Gel electrolyte

Organic-metallic hybrid polymer
(Electrochromic layer)

different displays by reversing the direction of the current. The device equipped with two **Ru/FeL1** films (**Ru/FeL1** is a copolymer of **RuL1** and **FeL1** bearing 1,4-bis(2,2′:6′,2″-terpyridin-4-yl)benzene as the organic ligand) therefore had five different types of display that could be activated by changing the applied potentials over the range of −3 and 3 V, because the **Ru/FeL1** films exhibited different colors such as red, orange, and pale green at potentials of 0, 1.8, and 2.5 V, respectively [21, 22].

Acknowledgments   This research was financially supported by JST-PRESTO and JST-CREST projects.

References

1. Comiskey B, Albert JD, Yoshizawa H, Jacobson J (1998) An electrophoretic ink for all-printed reflective electronic displays. Nature 394:253–255. doi:10.1038/28349
2. Chen Y, Au J, Kazlas P, Ritenour A, Gates H, McCreary M (2003) Electronic paper: flexible active-matrix electronic ink display. Nature 423:136. doi:10.1038/423136a
3. Gelinck GH, Huitema HEA, Van Veenendaal E, Cantatore E, Schrijnemakers L, Van der Putten J, Geuns TCT, Beenhakkers M, Giesbers JB, Huisman BH, Meijer EJ, Benito EM, Touwslager FJ, Marsman AW, Van Rens BJE, De Leeuw DM (2004) Flexible active-matrix displays and shift registers based on solution-processed organic transistors. Nat Mater 3:106–110. doi:10.1038/nmat1061
4. Rosseinsky DR, Mortimer RJ (2001) Electrochromic systems and the prospects for devices. Adv Mater 13:783–793. doi:10.1002/1521-4095
5. Groenendaal L, Zotti G, Aubert PH, Waybright SM, Reynolds JR (2003) Electrochemistry of poly(3,4-alkylenedioxythiophene) derivatives. Adv Mater 15:855–879. doi:10.1002/adma.200300376
6. Kolb U, Buscher K, Helm CA, Lindner A, Thunemann AF, Menzel M, Higuchi M, Kurth DG (2006) The solid-state architecture of a metallosupramolecular polyelectrolyte. Proc Natl Acad Sci U S A 103:10202–10206. doi:10.1073/pnas.0601092103
7. Kurth DG, Higuchi M (2006) Transition metal ions: weak links for strong polymers. Soft Matter 2:915–927. doi:10.1039/b607485e
8. Han FS, Higuchi M, Kurth DG (2007) Diverse synthesis of novel bisterpyridines via Suzuki-type cross-coupling. Org Lett 9:559–562. doi:10.1021/ol062788h
9. Han FS, Higuchi M, Ikeda T, Negishi Y, Tsukuda T, Kurth DG (2008) Luminescence properties of metallo-supramolecular coordination polymers assembled from pyridine ring functionalized ditopic bis-terpyridines and Ru(II) ion. J Mater Chem 18:4555–4560. doi:10.1039/b806930a
10. Han FS, Higuchi M, Kurth DG (2008) Synthesis of π-conjugated pyridine ring functionalized bis-terpyridines with efficient green, blue and purple emission. Tetraherdon 64:9108–9116. doi:10.1016/j.tet.2008.06.106
11. Pal RR, Higuchi M, Negishi Y, Tsukuda T, Kurth DG (2010) Fluorescent Fe(II) metallo-supramolecular polymers: metal-ion-directed self-assembly of new bisterpyridines containing triethylene glycol chains. Polym J 42:336–341. doi:10.1038/pj.2010.3
12. Bandyopadhyay A, Sahu S, Higuchi M (2011) Tuning of nonvolatile bipolar memristive switching in Co(III) polymer with an extended azo aromatic ligand. J Am Chem Soc 133:1168–1171. doi:10.1021/ja106945v
13. Li J, Futera Z, Li H, Tateyama Y, Higuchi M (2011) Conjugation of organic-metallic hybrid polymers and calf-thymus DNA. Phys Chem Chem Phys 13:4839–4841. doi:10.1039/c0cp02037k

14. Sato T, Higuchi M (2012) A vapoluminescent Eu-based metallo-supramolecular polymer. Chem Commun 48:4947–4949. doi:10.1039/c2cc30972f
15. Li J, Murakami T, Higuchi M (2013) Metallo-supramolecular polymers: versatile DNA binding and their cytotoxicity. J Inorg Organomet Polym Mater 23:119–125. doi:10.1007/s10904-012-9752-2
16. Hofmeier H, Schubert US (2004) Recent developments in the supramolecular chemistry of terpyridine-metal complexes. Chem Soc Rev 33:373–399. doi:10.1039/b400653b
17. Higuchi M, Kurth DG (2007) Electrochemical functions of metallo-supramolecular nano-materials. Chem Rec 7:203–209. doi:10.1002/tcr.20118
18. Han FS, Higuchi M, Kurth DG (2007) Metallo-supramolecular polymers based on function-alized bis-terpyridines as novel electrochromic materials. Adv Mater 19:3928–3931. doi:10.1002/adma.200700931
19. Han FS, Higuchi M, Kurth DG (2008) Metallo-supramolecular polyelectrolytes self-assembled from various pyridine ring substituted bis-terpyridines and metal ions: photo-physical, electrochemical and electrochromic properties. J Am Chem Soc 130:2073–2081. doi:10.1021/ja710380a
20. Hossain MD, Sato T, Higuchi M (2013) Green color copper-based metallo-supramolecu-lar polymer: synthesis, structure, and electrochromic properties. Chem Asian J 8:76–79. doi:10.1002/asia.201200668
21. Higuchi M, Akasaka Y, Ikeda T, Hayashi A, Kurth DG (2009) Electrochromic solid-state devices using organic-metallic hybrid polymers. J Inorg Organomet Polym Mater 19:74–78. doi:10.1007/s10904-008-9243-7
22. Higuchi M (2009) Electrochromic organic–metallic hybrid polymers: fundamentals and device applications. Polym J 41:511–520. doi:10.1295/polymj.PJ2009053

Chapter 3
Precisely Controlled Metal Nanoclusters

Yuichi Negishi

Abstract There have been remarkable advances in the development of gold nano-clusters protected by thiolates ($Au_n(SR)_m$), and techniques for the synthesis and characterization of these materials have improved significantly, enabling $Au_n(SR)_m$ to be synthesized with atomic precision. Experiments on the stabilities of clusters synthesized in this way have revealed a series of magic clusters, and the structures and physical and chemical properties of these magic clusters have subsequently been elucidated. Furthermore, several methods have been established for the functionalization of magic clusters. This chapter describes recent developments in $Au_n(SR)_m$ cluster chemistry.

Keywords Metal clusters • Precise synthesis • Magic cluster • Physical and chemical properties • Functionalization

3.1 Introduction

Small metal clusters (<2 nm) exhibit size-specific physical and chemical properties that are not observed in the corresponding bulk metals. Metal clusters have been studied for many years in a variety of different fields, including colloid chemistry and complex chemistry. The clusters available for research, however, are not very stable, which leads to difficulties in the handling and manipulation of these materials. In 1994, Brust et al. [1] reported the synthesis of thiolate-protected gold clusters ($Au_n(SR)_m$). The formation of a strong bond between the gold and the thiolate ligands in the clusters increased the stability of the $Au_n(SR)_m$ materials and allowed it

Y. Negishi (✉)
Department of Applied Chemistry, Faculty of Science, Tokyo University
of Science, Tokyo, Japan
e-mail: negishi@rs.kagu.tus.ac.jp

Y. Matsuo et al. (eds.), *Metal–Molecular Assembly for Functional Materials*,
SpringerBriefs in Molecular Science, DOI: 10.1007/978-4-431-54370-1_3,

to be treated as a single chemical substance. Following on from this report, $Au_n(SR)_m$ clusters have been investigated on the basis of their potential applications in a wide range of fields. These investigations have resulted in the development of a series of methodologies for the synthesis of clusters, as well as a deeper understanding of the basic properties of these materials, and techniques capable of increasing their functionality. This chapter describes recent developments in $Au_n(SR)_m$ cluster chemistry.

3.2 Precise Synthesis

$Au_n(SR)_m$ clusters are typically prepared in solution by reducing the polymers formed from the reactions between the gold ions and the thiols. There are, however, several disadvantages associated with this type of procedure, in that it yields a mixture of metal clusters with a variety of different chemical compositions (Fig. 3.1). To obtain individual $Au_n(SR)_m$ clusters, it is therefore necessary to separate the individual clusters at high resolution. Several procedures have been used as effective methods of separation for this purpose, including high-performance liquid chromatography and polyacrylamide gel electrophoresis (Fig. 3.1) [2–5]. Using an innovative electrophoresis approach, the separation of glutathione (GSH)-protected gold clusters ($Au_n(SG)_m$) was successfully accomplished at high resolution [6, 7]. The chemical compositions of the separated $Au_n(SG)_m$ clusters were accurately determined using mass spectrometry (Fig. 3.1). A series of similar experiments allowed for the isolation of several metal clusters, including $Au_{10}(SG)_{10}$, $Au_{15}(SG)_{13}$, $Au_{18}(SG)_{14}$, $Au_{22}(SG)_{16}$, $Au_{22}(SG)_{17}$, $Au_{25}(SG)_{18}$,

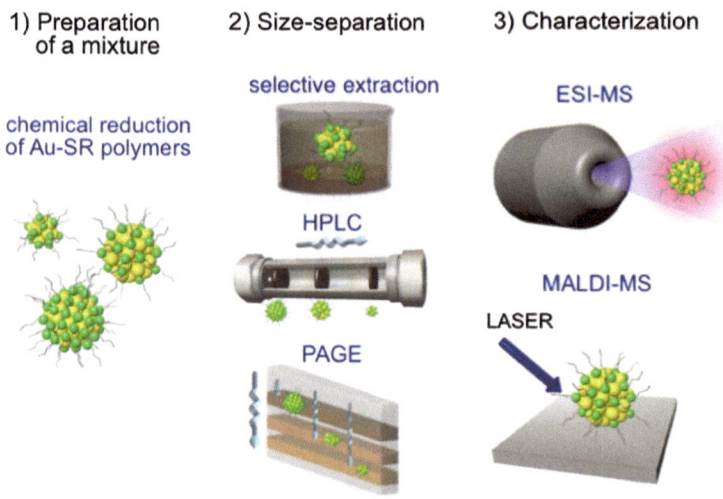

Fig. 3.1 Experimental flow chart for the precise synthesis of $Au_n(SR)_m$ clusters

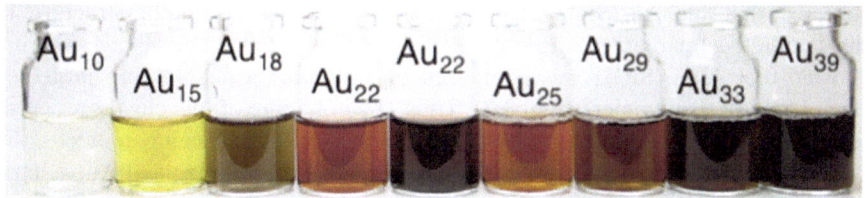

Fig. 3.2 Photograph of the aqueous solutions of the isolated $Au_n(SG)_m$ clusters [7]

$Au_{29}(SG)_{20}$, $Au_{33}(SG)_{22}$, and $Au_{39}(SG)_{24}$, in high purity (Fig. 3.2). These studies effectively demonstrated for the first time that $Au_n(SR)_m$ clusters could be treated as compounds with a well-defined chemical composition.

3.3 Magic Clusters

Although stability studies revealed that most of these $Au_n(SG)_m$ clusters were metastable species [7], only the $Au_{25}(SG)_{18}$ cluster was found to be both thermo-dynamically and chemically stable, and the cluster subsequently became known as the magic cluster [7, 8]. The synthesis of this magic cluster was found to be independent of the structure of the ligands [9, 10]. Similar experiments later revealed that $Au_{38}(SR)_{24}$ and $Au_{144}(SR)_{60}$ are also magic clusters [11]. More recently, a variety of other clusters have also been identified as magic clusters, including $Au_{20}(SR)_{16}$, $Au_{67}(SR)_{35}$, $Au_{102}(SR)_{44}$, $Au_{130}(SR)_{50}$, $Au_{187}(SR)_{68}$, and $Au_{333}(SR)_{79}$. Several studies have been reported in the literature showing that the production efficiency of these magic clusters is dependent on the synthetic conditions used in their production, and that the optimization of these conditions allows for the size-selective synthesis of clusters with a high atomic resolution [5, 12, 13].

The geometric structures of these $Au_n(SR)_m$ clusters were initially believed to be composed of a framework with a gold core protected by a group of simple thiolate units (Fig. 3.1). However, in 2007, Kornberg et al. [14] performed a single-crystal X-ray analysis of the structure of $Au_{102}(SR)_{44}$ (Fig. 3.3) which

Fig. 3.3 Geometric structures of $Au_{25}(SR)_{18}$ [15, 16], $Au_{38}(SR)_{24}$ [17], and $Au_{102}(SR)_{44}$ [14], which were elucidated by single-crystal X-ray analysis. For clarity, the R moieties are not shown

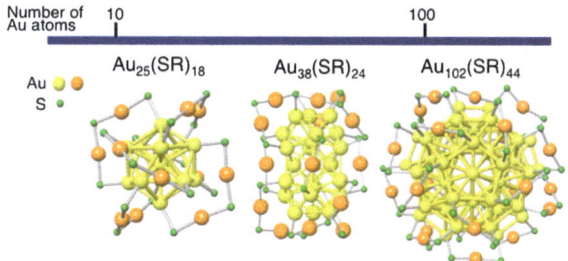

revealed that the $Au_n(SR)_m$ clusters possessed a very different structure to that predicted for them in the early literature in this field. The X-ray structure showed that $Au_{102}(SR)_{44}$ was composed of a symmetrical gold core protected by gold-thiolate oligomers (Fig. 3.3). Thus, the gold core of this cluster was protected by "gold-thiolate oligomers," as opposed to simply being protected by "thiolates". Following on from this report, $Au_{25}(SR)_{18}$ [15, 16] and $Au_{38}(SR)_{24}$ [17] (Fig. 3.3) were also found to be protected by similar oligomers. To date, only the geometric structures of these magic clusters have been elucidated. Interestingly, however, these $Au_n(SR)_m$ clusters have the gold-thiolate-ligand structure in common, and it is therefore likely that similar oligomers are formed in the other magic clusters [18].

The stability of these small magic clusters can be explained in terms of their geometrical and electronic factors. The $Au_{25}(SR)_{18}$, $Au_{38}(SR)_{24}$, and $Au_{102}(SR)_{44}$ clusters have a symmetrical metal core (Fig. 3.3). In addition, the total number of their valence electrons [19] satisfies the electronic structure of a closed shell [20]. These factors explain the stability of $Au_{25}(SR)_{18}$, $Au_{38}(SR)_{24}$, and $Au_{102}(SR)_{44}$ in terms of their geometry and electronic structure. In contrast, the total number of valance electrons in large clusters such as $Au_{130}(SR)_{50}$, $Au_{144}(SR)_{60}$, and $Au_{187}(SR)_{68}$ [19] does not satisfy the electronic structure of a closed shell. As the cluster size increases, the electron levels ultimately become continuous, and the cluster therefore ceases to have an electronic shell structure. This effect begins to appear in cluster sizes of $Au_{130}(SR)_{50}$ or larger, and it is envisaged that the stabilization of these clusters is derived from the geometric factors alone. Thus, for $Au_n(SR)_m$ clusters in general, as the cluster size approaches or exceeds the size of the $Au_{130}(SR)_{50}$ cluster, the origin of the stability of the cluster changes and shifts toward being geometric in nature [20].

Many studies have been performed on the physical and chemical properties of isolated magic clusters. These studies have revealed that small clusters, such as $Au_{25}(SR)_{18}$ and $Au_{38}(SR)_{24}$, possess a variety of interesting properties, including photoluminescence, paramagnetism, redox behavior, and optical and catalytic activities [3–5, 21–24]. The properties of these clusters have never been observed in bulk gold and are therefore unique to these clusters. The sources of some of these properties remain undetermined, and further experiments and theoretical studies are therefore required for a greater understanding of these materials.

3.4 Functionalization of Magic Clusters

As well as efforts directed towards elucidating the physical and chemical properties of magic clusters, studies have also been conducted to produce metal clusters with specific features. Heteroatom doping studies have revealed that doping $Au_{25}(SR)_{18}$ with a single palladium [25] or platinum [26] atom (Fig. 3.4) can lead to an increase in the overall stability of the clusters. Doping with these elements not only increased the stability of the clusters but also altered the chemical

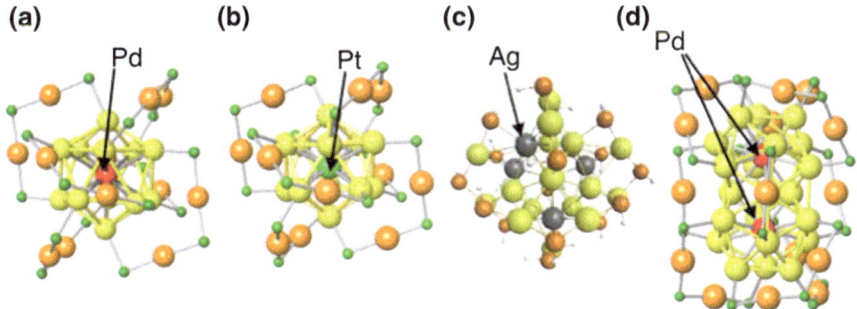

Fig. 3.4 Optimized structures for $Au_n(SR)_m$ clusters doped with a foreign atom: **a** $PdAu_{24}(SR)_{18}$ [25], **b** $PtAu_{24}(SR)_{18}$ [26], **c** $Ag_4Au_{21}(SR)_{18}$ [29], and **d** $Pd_2Au_{36}(SR)_{24}$ [31]. The R moieties have been omitted for clarity

properties, such as catalytic activity [26] and the reaction rate of ligand exchange reactions [27]. Doping with group 10 elements, such as silver [19, 28] or copper [30] (Fig. 3.4), changed the electronic structure and luminescence characteristics of the clusters. In addition, it has been suggested that doping with chromium, manganese, or iron would impart paramagnetic properties to the clusters. Foreign atom doping has also been applied to larger stable clusters, with the results revealing that foreign atom doping altered the stability and the electronic structure of $Au_{38}(SR)_{24}$ and $Au_{144}(SR)_{60}$ (Fig. 3.4) [31, 32]. It is envisaged that this will become widely used for the functionalization of $Au_n(SR)_m$ clusters because foreign atom doping modifies the electronic structure and physical properties of $Au_n(SR)_m$ clusters.

The protection afforded by functional thiols represents another effective method for the functionalization of $Au_n(SR)_m$ clusters. Photoresponsive $Au_{25}(SR)_{18}$ clusters have been produced by protection with azobenzene functionalized thiolates, which give rise to isomerization reactions upon photoirradiation (Fig. 3.5) [33]. In addition, the protection of the clusters with cyclodextrin,

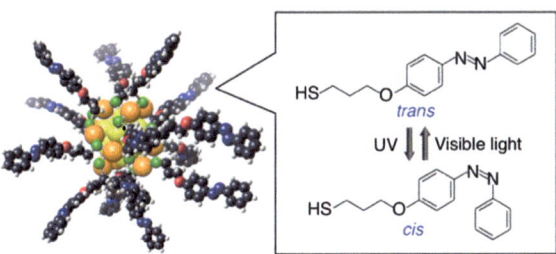

Fig. 3.5 Structure of an Au_{25} cluster protected by azobenzene derivative thiolates. This cluster alters the redox potential and optical absorption through the isomerization of azobenzenes in the ligands [33]. This structure was not optimized and is only shown for illustration purposes

which possesses molecular recognition abilities, produced gold clusters with similar recognition properties [34]. These studies showed that the chemical/physical properties of an $Au_n(SR)_m$ cluster could be controlled by adjusting the number of functional ligands incorporated into the cluster. Moreover, the ability to choose the number of functional ligands improved the ability to form a regular arrangement of $Au_n(SR)_m$ clusters and precisely place an $Au_n(SR)_m$ cluster between a series of electrodes. Unfortunately, however, techniques capable of providing control over the chemical composition of the ligands have not yet been developed for clusters containing two or more different types of ligands. It is generally expected that advanced control over the functional properties and the sequencing/arrangement of $Au_n(SR)_m$ clusters will be accomplished by further developments of techniques of this type [35].

The use of selenolates (SeR) as ligands is another effective method for the production of highly functional clusters. Compared with sulfur, selenium has electronegativity and atomic radius values closer to those of gold. Therefore, compared with Au–SR bonds, Au–SeR bonds possess a higher level of covalent character and greater bond energy. For these reasons, gold clusters synthesized using SeR ligands are more stable than the corresponding clusters synthesized using thiolate ligands. The SeR-protected Au_{25} clusters ($Au_{25}(SeR)_{18}$) [36] have been demonstrated to possess greater levels of stability towards degradation in solution and photodissociation than the corresponding $Au_{25}(SR)_{18}$ clusters [37]. The use of Se-type ligands also enabled the synthesis of alloy clusters that would be difficult to synthesize using the thiolates [38]. Thus, when SeR ligands are used, the clusters become more stable, and the synthesis of a variety of different alloys also becomes possible. In the future, $Au_n(SeR)_m$ clusters with SeR ligands should attract considerable levels of attention as new functional nanomaterials.

3.5 Summary and Perspective

In summary, recent research developments on $Au_n(SR)_m$ clusters have been described. Progress has been made in the development of $Au_n(SR)_m$ clusters, and techniques for the tailored synthesis of $Au_n(SR)_m$ clusters have consequently been established. In the future, $Au_n(SR)_m$ clusters should become commercially available, making them readily available for further study, and this should lead to a significant increase in the general use of $Au_n(SR)_m$ clusters. It is generally envisaged that additional studies in this area will result in the development of new synthetic methodologies and techniques for the creation of functional clusters. Developments in $Au_n(SR)_m$ cluster chemistry could potentially have a significant influence on the fields of materials science and nanotechnology.

Acknowledgments   The author would like to extend his sincerest thanks to Mr. Yoshiki Niihori and Mr. Wataru Kurashige (Tokyo University of Science) for their valuable comments.

References

1. Brust M, Walker M, Bethell D, Schiffrin DJ, Whyman R (1994) Synthesis of thiol-derivatised gold nanoparticles in a two-phase liquid–liquid system. J Chem Soc Chem Commun 801–802. doi:10.1039/c39940000801
2. Whetten RL, Khoury JT, Alvarez MM, Murthy S, Vezmar I, Wang ZL, Stephens PW, Cleveland CL, Luedtke WD, Landman U (1996) Nanocrystal gold molecules. Adv Mater 8:428–433. doi:10.1002/adma.19960080513
3. Parker JF, Fields-Zinna CA, Murray RW (2010) The story of a monodisperse gold nanoparticle: $Au_{25}L_{18}$. Acc Chem Res 43:1289–1296. doi:10.1021/ar100048c
4. Tsukuda T (2012) Toward an atomic-level understanding of size-specific properties of protected and stabilized gold clusters. Bull Chem Soc Jpn 85:151–168. doi:10.1246/bcsj.20110227
5. Jin R (2010) Quantum sized, thiolate-protected gold nanoclusters. Nanoscale 2:343–362. doi:10.1039/B9NR00160C
6. Negishi Y, Takasugi Y, Sato S, Yao H, Kimura K, Tsukuda T (2004) Magic-numbered Au_n clusters protected by glutathione monolayers ($n = 18, 21, 25, 28, 32, 39$): isolation and spectroscopic characterization. J Am Chem Soc 126:6518–6519. doi:10.1021/ja0483589
7. Negishi Y, Nobusada K, Tsukuda T (2005) Glutathione-protected gold clusters revisited: bridging the gap between gold(I)-thiolate complexes and thiolate-protected gold nanocrystals. J Am Chem Soc 127:5261–5270. doi:10.1021/ja042218h
8. Shichibu Y, Negishi Y, Tsunoyama H, Kanehara M, Teranishi T, Tsukuda T (2007) Extremely high stability of glutathionate-protected Au_{25} clusters against core etching. Small 3:835–839. doi:10.1002/amll.200600611
9. Negishi Y, Takasugi Y, Sato S, Yao H, Kimura K, Tsukuda T (2006) Kinetic stabilization of growing gold clusters by passivation with thiolates. J Phys Chem B 110:12218–12221. doi:10.1021/jp062140m
10. Negishi Y, Chaki NK, Shichibu Y, Whetten RL, Tsukuda T (2007) Origin of magic stability of thiolated gold clusters: a case study on $Au_{25}(SC_6H_{13})_{18}$. J Am Chem Soc 129:11322–11323. doi:10.1021/ja073580
11. Chaki NK, Negishi Y, Tsunoyama H, Shichibu Y, Tsukuda T (2008) Ubiquitous 8 and 29 kDa gold: alkanethiolate cluster compounds: mass spectrometric determination of molecular formulas and structural implications. J Am Chem Soc 130:8608–8610. doi:10.1021/ja8005379
12. Shichibu Y, Negishi Y, Tsukuda T, Teranishi T (2005) Large-scale synthesis of thiolated Au_{25} clusters via ligand exchange reactions of phosphine-stabilized Au_{11} clusters. J Am Chem Soc 127:13464–13465. doi:10.1021/ja053915s
13. Yu Y, Luo Z, Yu Y, Lee YJ, Xie J (2012) Observation of cluster size growth in CO-directed synthesis of $Au_{25}(SR)_{18}$ nanoclusters. ACS Nano 6:7920–7927. doi:10.1021/nn3023206
14. Kornberg RD, Jadzinsky PD, Calero G, Ackerson CJ, Bushnell DA (2007) Structure of a thiol monolayer-protected gold nanoparticle at 1.1 Å resolution. Science 318:430–433. doi:10.1126/science.1148624
15. Heaven MW, Dass A, White PS, Holt KM, Murray RW (2008) Crystal structure of the gold nanoparticle $[N(C_8H_{17})_4][Au_{25}(SCH_2CH_2Ph)_{18}]$. J Am Chem Soc 130:3754–3755. doi:10.1021/ja800561b
16. Zhu M, Aikens CM, Hollander FJ, Schatz GC, Jin R (2008) Correlating the crystal structure of a thiol-protected Au_{25} cluster and optical properties. J Am Chem Soc 130:5883–5885. doi:10.1021/ja801173r
17. Qian H, Eckenhoff WT, Zhu Y, Pintauer T, Jin R (2010) Total structure determination of thiolate-protected Au_{38} nanoparticles. J Am Chem Soc 132:8280–8281. doi:10.1021/ja103592z
18. De Jiang (2011) Staple fittness: a concept to understand and predict the structure of thiolated gold clusters. Chem Eur J 17:12289–12293. doi:10.1002/chem.201102391
19. Walter M, Akola J, Lopez-Acevedo O, Jadzinsky PD, Calero G, Ackerson CJ, Whetten RL, Grönbeck H, Häkkinen H (2008) A unified view of ligand-protected gold clusters as superatom complexes. Proc Natl Acad Sci U S A 105:9157–9162. doi:10.1073/pnas.0801001105

20. Negishi Y, Sakamoto C, Ohyama T, Tsukuda T (2012) Synthesis and the origin of the sta-
    bility of thiolate-protected Au_{130} and Au_{187} clusters. J Phys Chem Lett 3:1624–1628.
    doi:10.1021/jz300547d
21. Qian H, Zhu M, Wu Z, Jin R (2012) Quantum sized gold nanoclusters with atomioc preci-
    sion. Acc Chem Res 45:1470–1479. doi:10.1021/ar200331z
22. Dolamic I, Knoppe S, Dass A, Bürgi T (2012) First enantioseparation and circular dichroism
    spectra of Au_{38} clusters protected by achiral ligands. Nat Commun 3:798–803. doi:10.1038/
    ncomms1802
23. Sanchez-Castillo A, Noguez C, Garzon IL (2012) On the origin of the optical activity dis-
    played by chiral-ligand-protected metallic nanoclusters. J Am Chem Soc 132:1504–1505.
    doi:10.1021/ja907365f
24. Shibu ES, Muhammed MAH, Tsukuda T, Pradeep T (2008) Ligand exhange of Au25SG18
    leading to functionalized gold clusters: spectroscopy, kinetics, and luminescence. J Phys
    Chem C 112:12168–12176. doi:10.1021/jp800508d
25. Negishi Y, Kurashige W, Niihori Y, Iwasa T, Nobusada K (2010) Isolation, structure, and
    stability of a dodecanethiolate-protected Pd_1Au_{24} cluster. Phys Chem Chem Phys 12:6219–
    6225. doi:10.1039/b927175a
26. Qian H, De Jiang, Li G, Gayathri C, Das A, Gil RR, Jin R (2012) Monoplatinum dop-
    ing of gold nanoclusters and catalytic application. J Am Chem Soc 134:16159–16162.
    doi:10.1021/ja307657a
27. Niihori Y, Kurashige W, Matsuzaki M, Negishi Y (2013) Remarkable enhancement in ligand-
    exchange reactivity of thiolate-protected Au_{25} nanoclusters by single Pd atom doping.
    Nanoscale 5:508–512. doi:10.1039/c2nr32948d
28. Negishi Y, Iwai T, Ide M (2010) Continuous modulation of electronic structure of stable
    thiolate-protected Au_{25} cluster by Ag doping. Chem Commun 46:4713–4715. doi:10.1039/
    c0cc010121a
29. Guidez EB, Mäkinen V, Häkkinen H, Aikens CM (2012) Effects of silver doping on the
    geometric and electronic structure and optical absorption spectra of the $Au_{25-n}Ag_n(SH)^{18}$
    ($n = 1, 2, 4, 6, 8, 10, 12$) bimetallic nanoclusters. J Phys Chem C 116:20617–20624.
    doi:10.1021/jp306885u
30. Negishi Y, Munakata K, Ohgake W, Nobusada K (2012) Effect of copper doping on elec-
    tronic structure, geometric structure, and stability of thiolate-protected Au25 nanoclusters.
    J Phys Chem Lett 3:2209–2214. doi:10.1021/jz300892w
31. Negishi Y, Igarashi K, Munakata K, Ohgake W, Nobusada K (2012) Palladium doping of
    magic gold cluster $Au_{38}(SC_2H_4Ph)_{24}$: formation of $Pd_2Au_{36}(SC_2H_4Ph)_{24}$ with higher stabil-
    ity than $Au_{38}(SC_2H_4Ph)_{24}$. Chem Commun 48:660–662. doi:10.1039/c1cc15765e
32. Kumara C, Dass A (2012) AuAg alloy nanomolecules with 38 metal atoms. Nanoscale
    4:4084–4086. doi:10.1039/c2nr11781a
33. Negishi Y, Kamimura U, Ide M, Hirayama M (2012) A photoresponsive Au_{25} nanocluster pro-
    tected by azobenzene derivative thiolates. Nanoscale 4:4263–4268. doi:10.1039/c2nr30830d
34. Negishi Y, Tsunoyama H, Yanagimoto Y, Tsukuda T (2005) Subnanometer-sized gold clus-
    ters with dual molecular receptors: synthesis and assembly in one-dimensional arrangements.
    Chem Lett 34:1638–1639. doi:10.1246/cl.2005.1638
35. Niihori Y, Matzusaki M, Pradeep T, Negishi Y (2013) Separation of precise composi-
    tions of noble metal clusters protected with mixed ligands. J Am Chem Soc 135 (in press).
    doi:10.1021/ja4009369
36. Negishi Y, Kurashige W, Kamimura U (2011) Isolation and structural characterization of an
    octaneselenolate-protected Au_{25} cluster. Langmuir 27:12289–12292. doi:10.1021/la203301p
37. Kurashige W, Yamaguchi M, Nobusada K, Negishi Y (2012) Ligand-induced stabil-
    ity of gold nanoclusters: thiolate versus selenolate. J Phys Chem Lett 3:2649–2652.
    doi:10.1021/jz301191t
38. Kurashige W, Munakata K, Nobusada K, Negishi Y (2013) Synthesis of stable CunAu25-n
    nanoclusters ($n = 1$–9) using selenolate ligands. Chem Commun 49 (in press). doi:10.1039/
    C3CC41210E

Chapter 4
Coordination Nanocages for Engineering Discrete Aromatic Stacks

Michito Yoshizawa

Abstract The stacking of large aromatic molecules has been the subject of considerable attention, with the process imparting unique chemical and physical properties to the resulting systems. Although infinite aromatic stacks have been thoroughly explored, studies directed towards precisely controlled discrete assemblies composed of more than two aromatic molecules are relatively scarce. This chapter focuses on recent research towards the development of discrete stacks of large aromatic molecules. In contrast to conventional synthetic approaches, self-assembled coordination hosts with large box-shaped cavities act as useful molecular tools and facilitate the construction of discrete stacks of large aromatic molecules with unique properties.

Keywords Aromatic • Stack • Self-assembly • Coordination • Host

4.1 Introduction

Aromatic-aromatic interactions are fundamentally important non-covalent intermolecular forces that play an important role in determining the structure and properties of molecular assemblies. The stacking of large aromatic molecules is particularly attractive and imparts unique chemical and physical properties to the products where it occurs. For example, discotic liquid crystals are columnar assemblies of planar aromatic compounds that contain long side-chains, and organic electroconductive materials are constructed via the alternate charge-transfer stacking of electron-donating and -accepting extended aromatic compounds [1–3]. Although

M. Yoshizawa (✉)
Chemical Resources Laboratory, Tokyo Institute of Technology, Yokohama, Japan
e-mail: yoshizawa.m.ac@m.titech.ac.jp

Y. Matsuo et al. (eds.), *Metal–Molecular Assembly for Functional Materials*,
SpringerBriefs in Molecular Science, DOI: 10.1007/978-4-431-54370-1_4,
© The Author(s) 2013

infinite assemblies have been thoroughly explored, studies directed towards the construction of precisely controlled discrete assemblies composed of more than two aromatic molecules are relatively scarce. Discrete assemblies of a limited number of stacked molecules in a particular order are expected to show unique properties that would be different from those of the isolated or infinitely stacked systems. Given the potential significance of these systems, the development of a general method for the construction of these structures is required. Initial methods for preparing discrete stacks of aromatic molecules employed covalent linkers. Although these methods were impressive, the synthesis of covalent stacks was often lengthy and low yielding. Recently, supramolecular self-assembly has been established as a powerful method for the construction of nanoscale molecular architectures, and the focus has consequently shifted toward the utilizing of non-covalent interactions for the preparation of well-defined aromatic stacks [4, 5].

4.2 Box-Shaped Coordination Hosts

To avoid the lengthy covalent syntheses and purification processes associated with the existing methods, we employed box-shaped coordination hosts as molecular templates for the preparation of finite aromatic stacks. In 2005, we reported coordination host **1**, which consisted of a large box-shaped cavity that was capable of accommodating two planar molecule in a stacked fashion (Fig. 4.1) [6, 7]. Host **1** was quantitatively obtained from the self-assembly of three pillar ligands, two panel ligands, and six *cis*-capped Pd(II) complexes in the presence of templating aromatic molecules. Two aromatic molecules, such as pyrene, triphenylene, or coronene, could then form stacked dimers within the host through π-stacking and hydrophobic interactions.

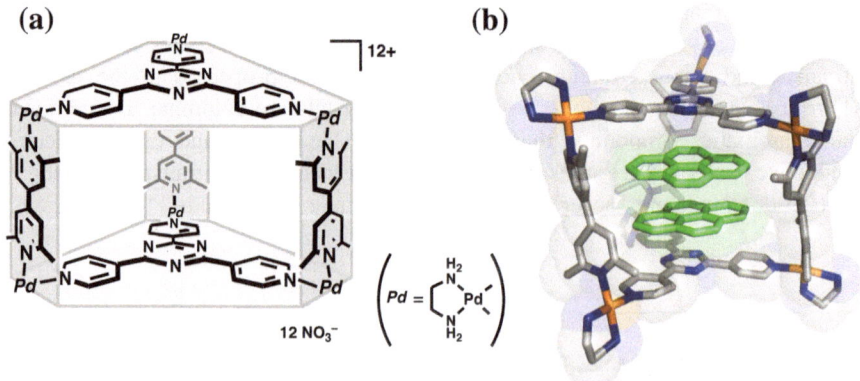

Fig. 4.1 **a** Box-shaped coordination host **1**. **b** X-ray crystal structure of **1** with two molecules of pyrene

4.3 Discrete Stacks of Planar Organic Molecules

Simply changing the length of the pillar ligand can lead to an increase in the cavity height and the number of stacked aromatics that can be placed within the cavity. Pyrene-4,5-dione (**2**) has a large dipole moment and forms infinite, head-to-tail columnar stacks in the solid state that effectively cancel the net dipole. When compound **2** was combined with the components of host **3** in an aqueous solution, the host–guest complex **3** ⊃ (**2**)₃ was quantitatively assembled and contained three stacked aromatic molecules [8]. X-ray crystallographic analysis revealed that the three guest molecules were not stacked in the usual head-to-tail fashion but that they were rotated by 120° with respect to each other (Fig. 4.2). By systematically altering the height of the cavity, discrete stacks composed of two, three, four, and five molecules of compound **2** could be constructed in a selective manner (Fig. 4.2c) [9].

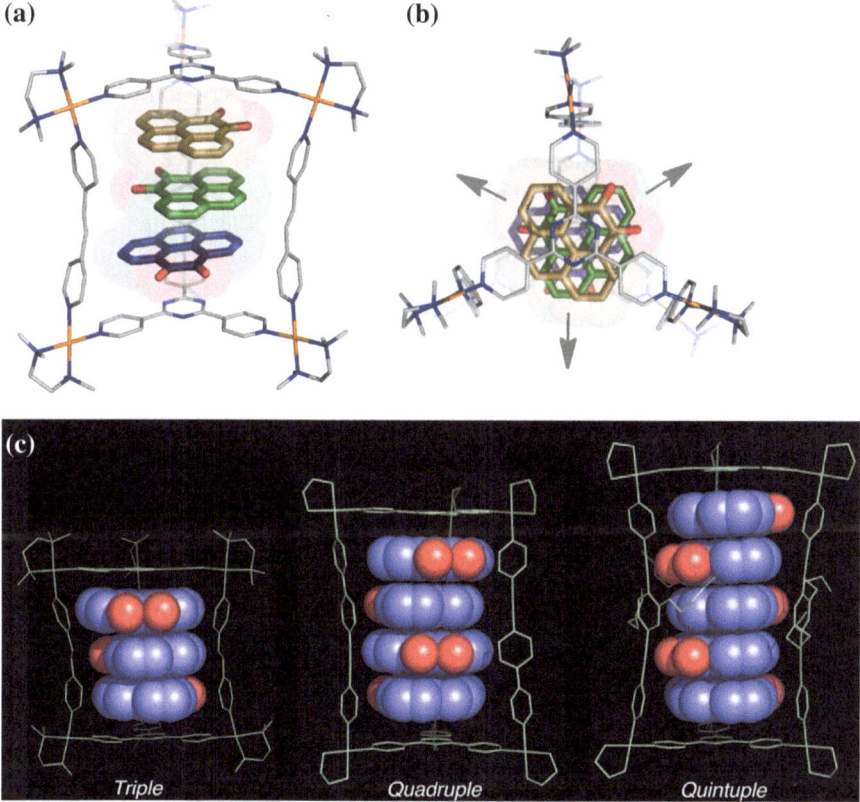

Fig. 4.2 X-ray crystal structure of triple stacked aromatic molecules **2** within host **3**: **a** the *side* and **b** *top* views. **c** Discrete stacks of three, four, and five molecules of planar **2** within coordination hosts

(a)

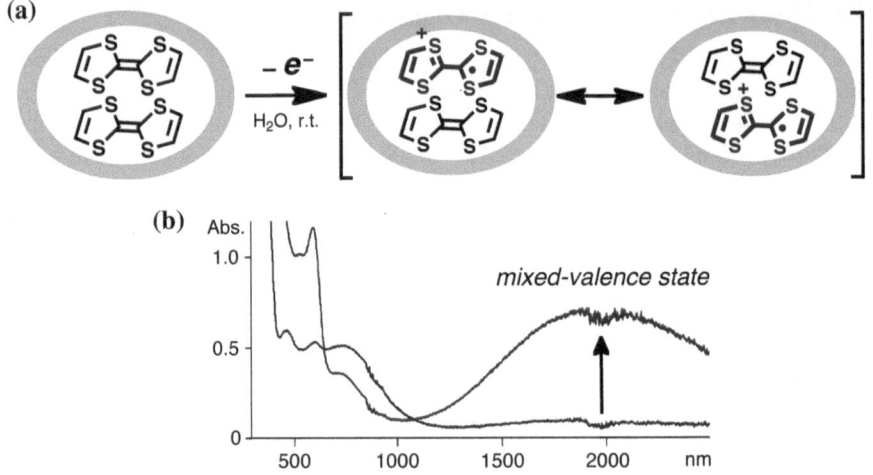

(b)

Fig. 4.3 **a** Stabilization of the mixed-valence state of stacked TTF dimer within host **1**. **b** UV–vis spectra of **1** ⊃ (TTF)$_2$ and **1** ⊃ [(TTF)$_2$]$^{\cdot+}$ at ambient temperature in an aqueous solution

Interestingly, an unusual mixed-valence state of a stacked tetrathiafulvalene (TTF) dimer was observed within the coordination host **1** at ambient temperature in an aqueous solution [10]. When excess TTF was added to an aqueous solution of **1**, the colorless solution became dark green in color because of the formation of a **1** ⊃ (TTF)$_2$ complex (Fig. 4.3a). Electrochemical studies revealed that an initial one-electron oxidation occurred at ~150 mV that led to the mixed-valence dimer, whereas a second one-electron oxidation at ~300 mV afforded the cation radical dimer. The mixed-valence state was also indicated by the appearance of a broad absorption band in the near-infrared region ($\lambda_{max} =$ ~2000 nm) of the UV–vis spectrum (Fig. 4.3b). The host framework effectively forced the two molecules of TTF into close proximity in the cavity. As a result, the labile mixed-valence dimer was protected from oxygen and solvent molecules.

4.4 Discrete Stacks of Planar Metal-Complexes

The exertion of precise control over the spatial arrangement of metal-complexes is of considerable interest for the development of molecular-based electronic and magnetic materials. With this in mind, our own work turned towards exploiting the stacked conformation of planar metal-complexes within box-shaped host **1** to generate specific metal–metal interactions. Following the addition of bisacetylacetonato metal-complexes **4** (M = Pt(II), Pd(II), and

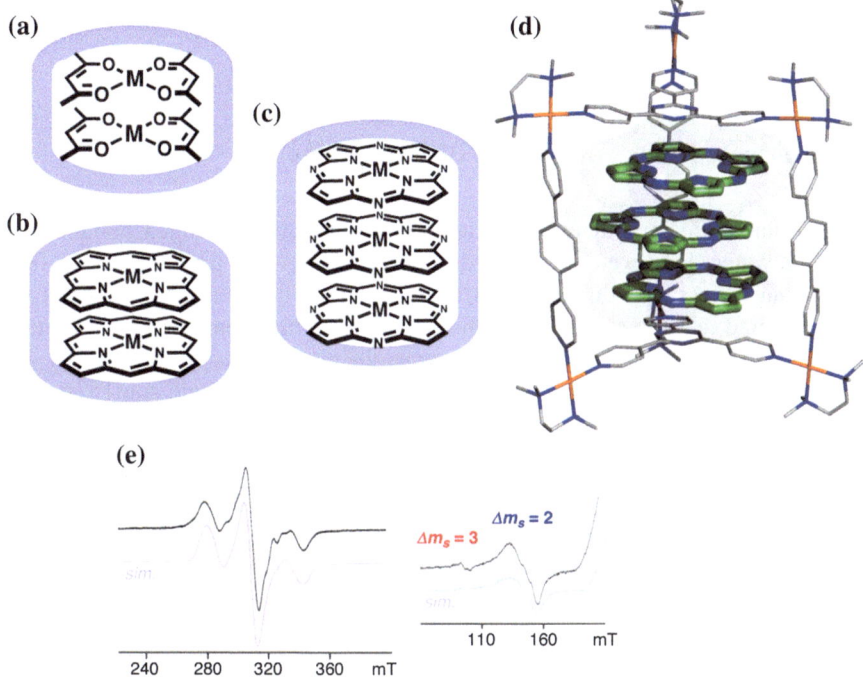

Fig. 4.4 Stacked dimers of **a** bisacetylacetonato metal-complexes **4** (M = Pt(II), Pd(II), and Cu(II)) and **b** metal-porphine molecules (M = Cu(II)) within coordination host **1**. **c** Triple stacks of metal-azaporphine molecules within box-shaped host **5** and **d** X-ray crystal structure of **5** ⊃ (**6**)$_3$ (M = H$_2$). **e** ESR spectra of **5** ⊃ (**6**)$_3$ (M = Cu(II)) at 103 K

Cu(II)) to an aqueous solution of **1**, two molecules of **4** entered the cavity and formed stable stacked dimers (Fig. 4.4a) [11, 12]. X-ray crystallographic analysis of the host–guest complex revealed that the Pt(II) atoms of the guests were in close contact (~3.3 Å). The metal–metal d–d interactions presented themselves in the UV–vis spectrum of the **1** ⊃ (**4**)$_2$ complex as broad absorptions (~450 nm).

The insertion of a phenylene spacer into the pillar ligand of **1** led to the extended coordination host **5**, which was suitable for accommodating triple stacks of azaporphine molecules (**6**). When Cu(II)-containing azaporphines were encapsulated in the host cavity, a Cu(II)-Cu(II)-Cu(II) array was obtained within the host (Fig. 4.4c, d) [13]. Exciton couplings between the stacked guests were observed in the UV–vis spectrum and ESR analysis revealed a ferromagnetically coupled quartet state ($S = 3/2$) for the three Cu(II) centers (Fig. 4.4e). The combination of electron-deficient metal-azaporphines and the electron-rich metal-porphines led to the formation of alternating mixed metal triple stacks such as Cu-Pd-Cu and Cu-Co–Cu [13]. Furthermore, the enclathration of two planar Ni(II)-complexes within host **1** was shown to induce unusual spin-crossover behavior [14].

4.5 Discrete Stacks of Planar Biomolecules

Short dinucleotide fragments are unable to form stable hydrogen-bonded base pairs in water. However, in the hydrophobic pockets of proteins, these fragments can form stable duplexes to transmit genetic information. Our work has demonstrated that the robust hydrophobic cavity of platinum host **7** can be used to stabilize and isolate minimal base pairs in an aqueous solution (Fig. 4.5) [15, 16]. The simple mixing of 5′-adenosine monophosphate and 5′-uridine monophosphate with **7** gave a 1:1:1 host–guest complex. X-ray crystallographic analysis of the complex clearly revealed the selective formation of a hydrogen-bonded A-U base pair (Fig. 4.5c). A platinum derivative possessed an expanded cavity and facilitated the assembly of an A-T base pair duplex in water (Fig. 4.5b).

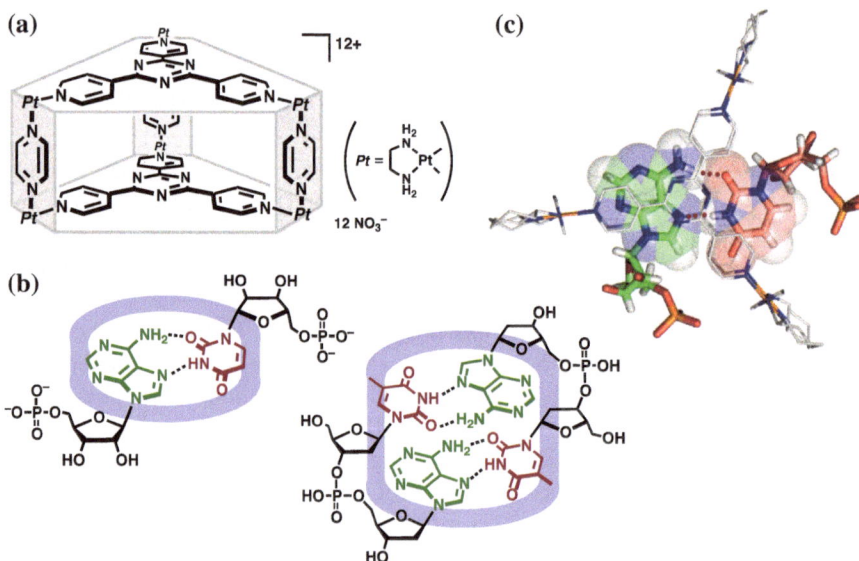

Fig. 4.5 **a** Box-shaped platinum host **7**. **b** The stabilization of the hydrogen-bonding pairs of mononucleotides and dinucleotides within the different host complexes. **c** X-ray crystal structure of a hydrogen-bonded A-U base pair within host **7**

4.6 Interlocked Aromatic Stacks

Inspired by the structures of catenanes [17], we envisaged that interlocked molecular hosts would provide a framework for the construction of extended stacked aromatic molecules in a discrete manner. When the appropriate pillar ligands, panel ligands, triphenylene molecules, and end-capped Pd(II) complexes were suspended in water in a 6:4:3:12 ratio, septuple aromatic stack **8** was obtained from the 25 components in quantitative yield (Fig. 4.6a) [18]. The ^{1}H NMR spectrum of the complex revealed two intercalated triphenylene signals in a 2:1 ratio, and both signals had been shifted upfield because of the shielding provided by the panel ligands (Fig. 4.6b). Analysis by mass spectrometry confirmed the expected molecular mass of **8**, and X-ray crystallographic analysis was used to provide unambiguously characterization of the structure of the complex. The 2.1 nm long aromatic column of **8** was formed through efficient aromatic–aromatic interactions between all seven of the aromatic subunits (Fig. 4.7a).

This method was then successfully applied to the construction of even taller discrete stacks. A series of elongated pillar ligands were designed and combined with different ratios of the panel ligands, metal ions, and pyrene guests to give the octuple and the nonuple stacks **9** and **10**, respectively, in almost quantitative yields. The X-ray crystal structure of **9** confirmed the 2.4 nm tall cylindrical stack of aromatic molecules and the even larger **10** was calculated to be 2.7 nm tall (Fig. 4.7b, c) [18].

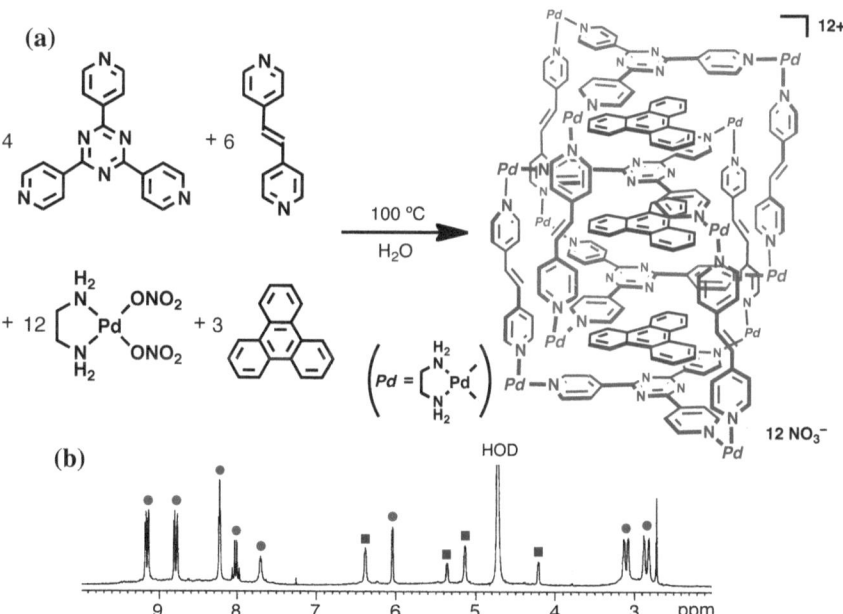

Fig. 4.6 a The quantitative formation of septuple aromatic stack **8** from the 25 components. **b** ^{1}H NMR spectrum of **8** in D$_2$O

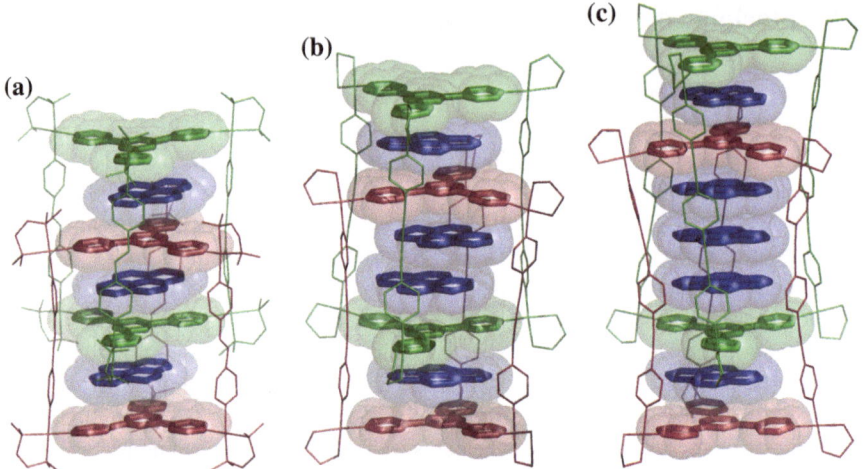

Fig. 4.7 X-ray crystal structures of (**a**) septuple stack **8** and (**b**) octuple stack **9**, and (**c**) the optimized structure of nonuple stack **10**

4.7 Infinite Stacks of Discrete Aromatic Stacks

Columnar stacks of aromatic molecules can generate unique physical properties in molecular assemblies. However, the number and variety of aromatic compounds suitable for stacking are rather limited. We envisaged that septuple aromatic stack **8** should stack and generate extended stacks with a height of $7 \times n$ (where n is the number of stacks). To enhance the solubility of these complexes and allow for further structural elaboration, the novel septuple aromatic stack **8′** was prepared using pillar ligands with solubilizing side chains. At increased concentrations (10-20 mM) in aqueous solutions, **8′** formed $7 \times n$ aggregates of the aromatic stacks through intermolecular hydrophobic and π-stacking interactions with $n \approx 4$ and 16, respectively (Fig. 4.8a) [19]. In spite of the modest degree of aggregation, impressive $7 \times n$ values of ~28 and ~112 were achieved in water. The simple addition of water to solid **8′** induced the formation of lyotropic liquid-crystalline mesophases (Fig. 4.8b), which could be potential applied as a tunable liquid-crystalline material.

4.8 Conclusions and Perspectives

This chapter has focused on recent efforts directed towards the development of discrete stacks of large aromatic molecules. In contrast to conventional synthetic chemistry, self-assembled coordination hosts have been prepared consisting of large box-shaped cavities and explored as useful tools for the construction of

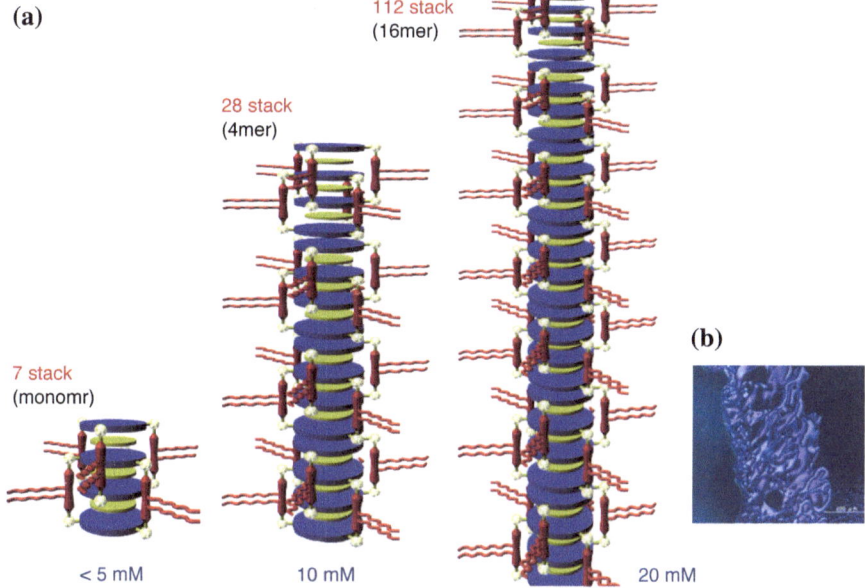

Fig. 4.8 **a** Schematic presentation of $7 \times n$ columnar stacks of aromatic stack **8'**. **b** Optical polarizing microscope texture of the columnar stacks of **8'** in H_2O

discrete stacks of large aromatic molecules with unique electronic and magnetic properties. The size and shape of the molecular box are tunable and precisely regulate the number and order of the stacked molecules. These simple yet powerful tools could provide a promising platform capable of engineering unique molecular elements for functional devices and materials [20–27].

Acknowledgments The author would like to extend his sincerest appreciation to all of his collaborators (especially Prof. Makoto Fujita) for their efforts and cooperation, as well as all of the researchers whose names are cited in the references. This project has been supported in part by grants from the Ministry of Education, Culture, Sports, Science and Technology of Japan.

References

1. Hunter CA (1994) Meldola lecture. The role of aromatic interactions in molecular recognition. Chem Soc Rev 23:101–109. doi:10.1039/CS9942300101
2. Hunter CA, Lawson KR, Perkins J, Urch CJ (2001) Aromatic interactions. J Chem Soc Perkin Trans 2:651–669. doi:10.1039/B008495F
3. Bushby RJ, Lozman OR (2002) Discotic liquid crystals 25 years on. Curr Opin Colloid Interface Sci 7:343–354. doi:10.1016/S1359-0294(02)0085-7
4. Yoshizawa M, Klosterman JK, Fujita M (2009) Functional molecular flasks: new properties and reactions within discrete, self-assembled hosts. Angew Chem Int Ed 48:3418–3438. doi: 10.1002/anie.200805340

5. Klosterman JK, Yamauchi Y, Fujita M (2009) Engineering discrete stacks of aromatic molecules. Chem Soc Rev 38:1714–1725. doi:10.1039/B901261N

6. Yoshizawa M, Nakagawa J, Kumazawa K, Nagao M, Kawano M, Ozeki T, Fujita M (2005) Discrete stacking of large aromatic molecules within organic-pillared coordination cages. Angew Chem Int Ed 44:1810–1813. doi:10.1002/anie.200462171

7. Yoshizawa M, Nagao M, Kumazawa K, Fujita M, Organomet J (2005) Side chain-directed complementary cis-coordination of two pyridines on Pd(II): selective multicomponent assembly of square-, rectangular-, and trigonal prism-shaped molecules. J Organomet Chem 690:5383–5388. doi:10.1016/j.jorganchem.2005.06.022

8. Yamauchi Y, Yoshizawa M, Akita M, Fujita M (2009) Discrete stack of an odd number of polarized aromatic compounds revealing the importance of net vs. local dipoles. Proc Natl Acad Sci U S A 106:10435–10437. doi:10.1073/pnas.0810319106

9. Yamauchi Y, Yoshizawa M, Akita M, Fujita M (2010) Engineering double to quintuple stacks of a polarized aromatic in confined cavities. J Am Chem Soc 132:960–966. doi:10.1021/ja904063r

10. Yoshizawa M, Kumazawa K, Fujita M (2005) Room-temperature and solution-state observation of the mixed-valence cation radical dimer of tetrathiafulvalene, $[(TTF)_2]^{+\bullet}$, within a self-assembled cage. J Am Chem Soc 127:13456–13457. doi:10.1021/ja053508g

11. Yoshizawa M, Ono K, Kumazawa K, Kato T, Fujita M (2005) Metal–Metal d–d interaction through the discrete stacking of mononuclear M(II) complexes (M = Pt, Pd, and Cu) within an organic-pillared coordination cage. J Am Chem Soc 127:10800–10801. doi:10.1021/ja053009f

12. Ono K, Yoshizawa M, Kato T, Watanabe K, Fujita M (2007) Porphine dimeric assemblies in organic-pillared coordination cages. Angew Chem Int Ed 46:1803–1806. doi:10.1002/anie.200604790

13. Ono K, Yoshizawa M, Kato T, Fujita M (2008) Three-metal-center spin interactions through the intercalation of metal azaporphines and porphines into an organic pillared coordination box. Chem Commun 2328–2330. doi:10.1039/B801701H

14. Ono K, Yoshizawa M, Akita M, Kato T, Tsunobuchi Y, Ohkoshi S, Fujita M (2009) Spin crossover by encapsulation. J Am Chem Soc 131:2782–2783. doi:10.1021/ja8089894

15. Sawada T, Yoshizawa M, Sato S, Fujita M (2009) Minimal nucleotide duplex formation in water through enclathration in self-assembled hosts. Nat Chem 1:53–56. doi:10.1038/nchem.100

16. Sawada T, Fujita M (2010) A single Watson-Crick G•C base pair in water: aqueous hydrogen bonds in hydrophobic cavities. J Am Chem Soc 132:7194–7201. doi:10.1021/ja101718c

17. Fujita M (1999) Self-assembly of [2]Catenanes containing metals in their backbones. Acc Chem Res 32:53–61. doi:10.1021/ar9701068

18. Yamauchi Y, Yoshizawa M, Fujita M (2008) Engineering stacks of aromatic rings by the interpenetration of self-assembled coordination cages. J Am Chem Soc 130:5832–5833. doi:10.1021/ja077783+

19. Yamauchi Y, Hanaoka Y, Yoshizawa M, Akita M, Ichikawa T, Yoshio M, Kato T, Fujita M (2010) $m \times n$ Stacks of discrete aromatic stacks in solution. J Am Chem Soc 132:9555–9557. doi:10.1021/ja103180z

20. Osuga T, Murase T, Ono K, Yamauchi Y, Fujita M (2010) m × n Metal Ion arrays templated by coordination cages. J Am Chem Soc 132:15553–15555. doi:10.1021/ja108367j

21. Kiguchi M, Takahashi T, Takahashi Y, Yamauchi Y, Murase T, Fujita M, Tada T, Watanabe S (2011) Electron transport through single molecules comprising aromatic stacks enclosed in self-assembled cages. Angew Chem Int Ed 50:5708–5711. doi:10.1002/anie.201100431

22. Kishi N, Li Z, Yoza K, Akita M, Yoshizawa M (2011) An M_2L_4 Molecular capsule with an anthracene shell: encapsulation of large guests up to 1 nm. J Am Chem Soc 133:11438–11441. doi:10.1021/ja2037029

23. Li Z, Kishi N, Yoza K, Akita M, Yoshizawa M (2012) Isostructural M_2L_4 molecular capsules with anthracene shells: synthesis, crystal structures, and fluorescent properties. Chem Eur J 18:8358–8365. doi:10.1002/chem.201200155

24. Yazaki K, Kishi N, Akita M, Yoshizawa M (2013) A Bowl-shaped organic host using bispyridine ligands: selective encapsulation of carbonyl guests in water. Chem Commun 49:1630–1632. doi:10.1039/C3CC38869G
25. Li Z, Ishizuka H, Sei Y, Akita M, Yoshizawa M (2012) Extended fluorochromism of anthracene trimers with a *meta*-substituted triphenylamine or triphenylphosphine core. Chem Asian J 7:1789–1794. doi:10.1002/asia.201200310
26. Kondo K, Suzuki A, Akita M, Yoshizawa M (2013) Micelle-like molecular capsules with anthracene shells as photoactive hosts. Angew Chem Int Ed 52:2308–2312. doi:10.1002/anie.201208643
27. Kishi N, Li Z, Akita M, Yoza K, Siegel JS, Yoshizawa M (2013) Wide-ranging host capability of a Pd(II)-linked M_2L_4 molecular capsule with an anthracene shell. Chem Eur J 19 (in press). doi:10.1002/chem.201204010

Chapter 5
Coordination Nanochannels for Polymer Materials

Takashi Uemura

Abstract Porous coordination polymers prepared via the self-assembly of metal ions and organic ligands have attracted considerable attention because of their potential applications in storage, separation, and catalytic systems. The use of their regulated nanochannels as the fields for polymerization can allow for precise control over the polymer structures. In addition, the confinement of polymer chains in the nanochannels allows for the formation of unique nanocomposites that show unprecedented and interesting dynamic, optical, and electronic properties.

Keywords Porous coordination polymers • Nanochannels • Polymerization • Nanoconfinement • Composites

5.1 Introduction

Recently, remarkable progress has been made in the area of porous coordination polymers (PCPs) prepared by the self-assembly of metal ions and bridging organic ligands, because of the diverse topologies of the polymers and their fascinating properties, as well as their potential application in several areas such as storage, exchange, and catalytic systems (Fig. 5.1) [1–3]. PCPs have unique characteristics, including framework regularity, high porosity, flexibility, and a tunable pore surface, which can be used to create high-performance pores, as well as unprecedented porous functionalities. Many investigations have been conducted over the past decade into the formation of highly porous and flexible frameworks.

T. Uemura (✉)
Department of Synthetic Chemistry and Biological Chemistry, Graduate School of Engineering, Kyoto University, Kyoto, Japan
e-mail: uemura@sbchem.kyoto-u.ac.jp

Y. Matsuo et al. (eds.), *Metal–Molecular Assembly for Functional Materials*,
SpringerBriefs in Molecular Science, DOI: 10.1007/978-4-431-54370-1_5,
© The Author(s) 2013

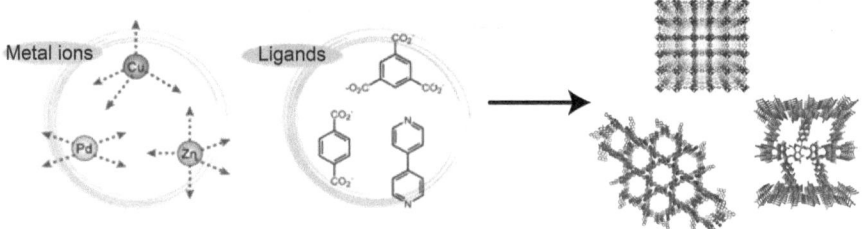

Fig. 5.1 Schematic illustration for the preparation of PCPs

The exertion of some level of control over the molecular reactions and transformations that can occur within the confined environments of the functional channels of PCPs is currently an important area of research, which has prompted vigorous efforts towards the development of PCPs for applications in heterogeneous catalysis, asymmetric reactions, and "ship-in-a-bottle" type syntheses [4, 5]. The use of PCP nanochannels in polymerization research represents a particularly attractive idea, which would allow for the exertion of multiple levels of control over the polymerization process (i.e., control over the stereo- and regioregularity, as well as control over the molecular weight, and polymer topology) as a consequence of efficient through-space induction mechanisms and specific host-monomer interactions [6, 7]. These functional nanochannels can be applied to the construction of tailor-made polymerization systems to obtain preferred polymer structures because of the highly designable features of PCPs. The preparation of host–guest nanocomposites based on PCPs and polymers is also of interest because of their host–guest synergistic and nanosize-dependent properties, as well as their application to the synthesis of nanomaterials.

This chapter will focus on discussing the recent progress and future perspectives of the wide variety of different polymerization systems that use PCPs for controlling primary and secondary polymer structures as well as obtaining specific properties based on low-dimensional assemblies of polymers and functional polymer-based nanohybrids. Several nanoporous matrices with regular nanochannel structures, such as supramolecular organic hosts, microporous zeolites, clays, and mesoporous materials, have played important roles in controlling the polymerization of the monomers accommodated in their channels structures by through-space interactions between the walls of the pores and the monomers [8–10]. There are, however, still many issues that need to be addressed and improved upon regarding pore size, stability, and surface functionality of these materials. For example, the occurrence of large strains or topological changes during the inclusion polymerization processes in organic hosts often results in the decomposition of the host channel structures, because the organic hosts are only held together by hydrogen bonding or weak van der Waals interactions. The mesopores are much larger than the conventional monomers, so the primary structures of polymers cannot be efficiently controlled in the mesopores. Thus, the design and preparation of new microporous materials with

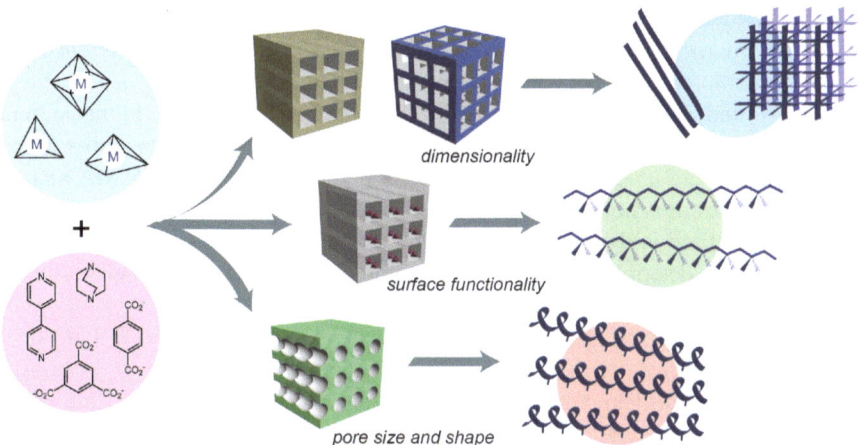

Fig. 5.2 Polymerization in PCP channels

designable nanospaces based on PCPs is highly desirable to allow for the development of precisely controlled polymerization processes that are applicable to a variety of different monomers (Fig. 5.2).

5.2 Controlled Radical Polymerization in PCPs

The radical polymerization of vinyl monomers was recently performed in the nanochannels of a variety of different PCPs of the general formula $[M_2(L)_2ted]_n$ ($M = Cu^{2+}$ or Zn^{2+}, L = dicarboxylate ligands, ted = triethylenediamine), and the relationships between the structures of the channels and the polymerization behaviors, including the monomer reactivity, molecular weight, and stereostructure, were studied [11–15]. The PCPs could be precisely tuned by changing the organic ligands, which allowed for the systematic study of the radical polymerization process in a system composed of nanoporous materials. During this particular polymerization, the propagation radicals at the ends of the polymer were remarkably stabilized by the efficient protection of the polymer chains in the nanochannels, which resulted in fewer unfavorable side reactions [11, 12]. The level of monomer conversion was typically reduced with a reduction in the pore size because of the lower mobility of the monomers [12].

A significant nanochannel effect was also observed on the stereoregularity of the polymer product, in that the tacticity of the polymers was found to be strongly dependent on the size and shape of the pores [12, 13]. For example, the polymerization of methyl methacrylate (MMA) in $[Cu_2(2,5\text{-dimethoxyterephthalate})_2(ted)]_n$ gave poly(methyl methacrylate) (PMMA) with high isotactic and heterotactic triad fractions (m = 54 %), and therefore represents one of the most effective systems for changing the tacticity of PMMA during a radical polymerization [13]. The stereocontrolled polymerization of MMA was also performed in a variety of different PCPs of the general formula [M(1,3,5-benzenetrisbenzoate)] (M $=$ Al^{3+}, Eu^{3+}, Nd^{3+}, Y^{3+}, La^{3+}, and Tb^{3+}), which contained unsaturated metal sites [14]. In this system, the composition of the isotactic units in PMMA increased when hosts with a higher Lewis acidity were used because of the effective interaction between MMA and the unsaturated sites. It is noteworthy that a PCP composed of Tb^{3+} gave PMMA with a significant increase in the composition of the isotactic units, although discrete Tb^{3+} complexes were determined to be ineffective for changing the stereoregularity of the PMMA in the solution polymerization system.

The radical polymerization reactions of divinylbenzenes (DVBs) in bulk or in solution typically results in the formation of hyper-branched network polymers, because the reactivities of the two vinyl moieties in the DVBs are equivalent. Interestingly, however, the radical polymerization reaction of DVBs in the one-dimensional channels of [M$_2$(terephthalate)$_2$ted]$_n$ (M $=$ Cu^{2+} or Zn^{2+}) gave a linear polymer with the pendant vinyl groups remaining unchanged in all of the benzene rings [15]. PCPs could therefore be used to successfully direct the linear polymerization of multivinyl monomers by effectively enveloping the reactive propagating radical mediators in a one-dimensional nanochannels.

5.3 Catalytic Polymerizations in PCPs

Compared with the conventional porous materials, one of the intriguing features of PCPs is that the functionality on the surfaces of their pores can be rationally controlled by the design of the constituents. The versatile pore features (e.g., redox activity, Lewis acidity, basicity, hydrophobicity, and chirality) of PCPs are of critical importance for the construction of unique nanosize reaction fields based on the PCP materials. It is therefore envisaged that functionalized PCPs with activation sites within the nanochannels could be applied in a variety of different catalytic applications in organic and polymer syntheses [4, 5].

The appropriate arrangement of interactive catalytic sites within the nanochannels of PCPs can provide remarkable nanochannel effects on the polymerization of substituted acetylenes [16]. For example, acidic methyl propiolate (MP) can be encapsulated in the nanochannels of [Cu$_2$(pyrazine-2,3-dicarboxylate)$_2$ (4,4′-bipyridine)]$_n$ and subsequently interact with the basic carboxylate oxygen atoms on the pore surface. C–H bond dissociation of the MP can then occur as a consequence of this acid–base interaction, leading to the production of a reactive acetylide species that initiates polymerization. In this work, the narrow nanochannels of the PCP significantly accelerated the polymerization process and successfully directed the *trans*-selective polymerization [16]. This result occurred in striking contrast to that obtained in the reaction using a discrete model catalyst, where an unfavorable mixture of the cyclic trisubstituted benzenes (major product) and the lesser favored conjugated polymer with a *cis* geometry (minor product) was obtained.

Many of the known PCPs have the potential to provide nanospaces with regularly arranged and highly dense redox-active sites. These nanospaces could be used to generate novel materials not only for highly selective catalysts and reaction fields, but also for the controlled orientation of the resulting products within the composites. The oxidative polymerization of pyrrole within the host layers of a PCP with $[Fe^{III}(CN)_6]^{3-}$ units was recently reported [17]. In this system, the host framework induced the oxidative polymerization of the pyrrole, whilst maintaining its crystallinity and morphology, which resulted in the formation of a layer-by-layer-type nanocomposite material. The removal of the host framework in an ethylenediamine-tetraacetic acid solution allowed for the isolation of the intercalated PPy as an insoluble black precipitate. Interestingly, microscopic analysis of the isolated polypyrrole revealed an extended 2-D sheet-stacking morphology, indicating that the isolated intercalated chains were isolated had maintained their orientation. The resulting microplate exhibited anisotropic electric conductivity properties, which resulted from the controlled orientation of the polymer chains in the plate object. In addition to the 2-D arrangement, a feasible method has been reported for the preparation of porous PPy using the 3-D nanochannels of a PCP as a template [18]. It is noteworthy that the porosity of the PPy can be controlled by changing the preparation conditions. Significant changes in the porous properties of the specific PPy structures were observed when oxygen or solvent molecules were used as the adsorbate.

5.4 Polymer Confinement in PCPs

Polymers confined in the nanosized spaces of the PCP channels typically show properties that are distinctly different from those shown for the same materials in the bulk state because of the formation of specific molecular assemblies and conformations [19, 20]. The inclusion of polymers within crystalline microporous hosts (pore size < 2 nm) with ordered and well-defined nanochannel structures has attracted considerable levels of attention because, in contrast to amorphous bulk polymer systems and polymers in solution, this approach can prevent the entanglement of polymer chains and provide extended chains in restricted spaces.

The inclusion of polymers can be achieved in the nanochannels of PCPs, which can lead to a fundamental study on the properties of the confined polymer chains (Fig. 5.3). A study was recently reported on the conformation and dynamics of single polystyrene (PSt) chains confined in $[Zn_2(terephthalate)_2ted]_n$ [21]. The PSt that was accommodated in regular 1-D nanochannels showed unique molecular dynamics. Solid-state NMR measurement clearly showed that the bulk PSt provides a wide motional distribution that originated from the heterogeneous local environment of the phenyl rings. In contrast, however, the PSt in the channel represented a quasi-single-type phenyl flipping motion, indicating the mobility of the homogeneous side-chain in the nanochannel. In addition, the activation energy for the phenyl flip of the nanoconfined PSt was significantly lower than that of the bulk material.

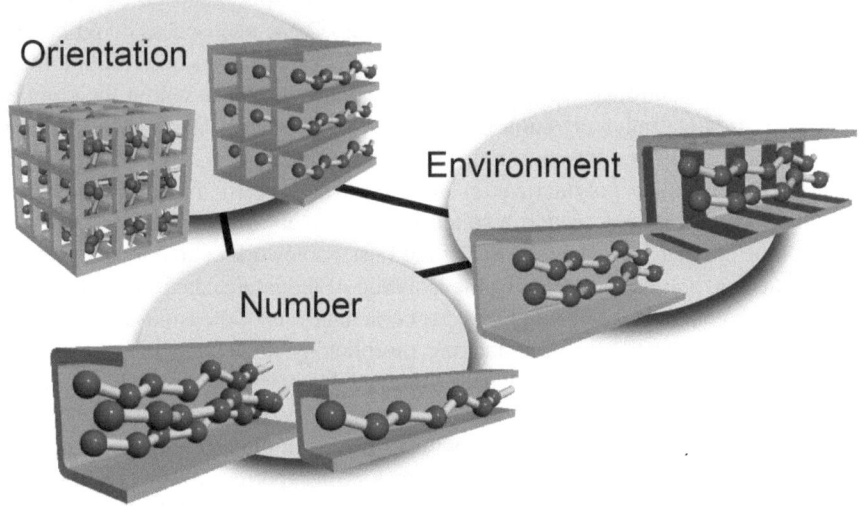

Fig. 5.3 Controlled the confinement of polymer chains within the PCP channels

An understanding of the thermal transitions of polymers at the level of a single-chain or few-chains is of considerable importance from the perspective of the future applications of these materials as nanosized transporters, lubricants, and connectors in molecular-based devices. Unfortunately, however, the thermal transitions of such ultrathin polymer assemblies, which exist below the 1 nm region, have remained obscure. The incorporation of poly(ethylene glycol) (PEG) into the regular nanochannels of PCPs $[M_2(L)_2ted]_n$ ($M = Cu^{2+}$ or Zn^{2+}, L = dicarboxylate ligands) has allowed for the thermal transition (i.e., melting-freezing) of the chain assemblies to be observed by differential scanning calorimetry, even in sub-nanometer pores [22]. The pore size (0.4–1.1 nm) and surface functionality of PCPs can be readily tailored by changing the ligand L, which allows for the transition behavior of the confined PEG to be systematically controlled. In this system, the reduction of transition temperature can be understood in terms of the destabilization of the solid state in confined spaces. The elevation of the transition temperature of PEG can be attributed to the interactions between the guest polymers with the hosts at their interface. Therefore, the molecular dynamics of PEG in PCPs can be determined by counterbalancing the size effects and the pore–guest interactions. It is noteworthy that the transition temperature of the confined PEG decreased as the molecular weight of the PEG increased. The introduction of Li ions into the PCP–PEG composite led to the liquid-like high mobility of the Li, which is particularly important for lithium ion batteries and the development of any future nanoelectronic applications [23].

PCPs have emerged as particularly exciting solid materials that effectively couple the features of crystallinity and flexibility. The construction of a unique host–guest cooperative system via the hybridization of a flexible PCP, $[Zn_2(tere-phthalate)_2ted]_n$, with a fluorescent guest oligomer, distyrylbenzene (DSB), was recently reported in the literature [24]. The PCP–DSB hybrid material was capable of adsorbing vapor and gas molecules at specific pressures. These adsorption phenomena were accompanied by structural transformations of the host PCP, which were coupled with conformational changes of the guest DSB. Accordingly, the guest fluorescence was changed significantly at specific vapor/gas pressures, and the selective fluorescence sensing of a particular gas was also achieved. These cooperative functions of the host–guest hybrid system represent a powerful concept for creating stimuli-responsive smart materials.

5.5 Conclusion

The use of regulated and tunable channels within PCPs for polymerization reactions could allow for multiple levels of control over the structures of the resulting polymers. In addition, the construction of nanocomposites between PCPs and polymers could provide access to unprecedented material platforms that would be capable of accomplishing a wide range of nanoscale functions. There are still many fundamental and fascinating aspects of polymer chemistry that could be performed in the nanochannels of PCPs.

References

1. Yaghi OM, O'Keeffe M, Ockwig NW, Chae HK, Eddaoudi M, Kim J (2003) Nature 423:705–714
2. Bradshaw D, Claridge JB, Cussen EJ, Prior TJ, Rosseinsky MJ (2005) Acc Chem Res 38:273–282
3. Ferey G, Serre C (2009) Chem Soc Rev 38:1380–1399
4. Corma A, Garcia H, Llabres i Xamena FX (2010) Chem Rev 110:4606–4655
5. Lee J, Farha OK, Roberts J, Scheidt KA, Nguyen ST, Hupp JT (2009) Chem Soc Rev 38:1450–1459
6. Uemura T, Horike S, Kitagawa S (2006) Chem Asian J 1:36–44
7. Uemura T, Yanai N, Kitagawa S (2009) Chem Soc Rev 38:1228–1236
8. Tajima K, Aida T (2000) Chem Commun 2399–2412
9. Miyata M (1996) Comprehensive supramolecular chemistry, vol 10. Pergamon, Oxford, pp 557–582
10. Moller K, Bein T (1998) Chem Mater 10:2950–2963
11. Uemura T, Kiatagwa K, Horike S, Kawamura T, Kitagawa S, Mizuno M, Endo K (2005) Chem Commun 5968–5970
12. Uemura T, Ono Y, Kitagawa K, Kitagawa S (2008) Macromolecules 41:87–94
13. Uemura T, Ono Y, Hijikata Y, Kitagawa S (2010) J Am Chem Soc 132:4917–4924
14. Uemura T, Uchida N, Higuchi M, Kitagawa S (2011) Macromolecules 44:2693–2697
15. Uemura T, Hiramatsu D, Kubota Y, Takata M, Kitagawa S (2007) Angew Chem Int Ed 46:4987–4990
16. Uemura T, Kitaura R, Ohta Y, Nagaoka M, Kitagawa S (2006) Angew Chem Int Ed 45:4112–4116
17. Yanai N, Uemura T, Ohba M, Kadowaki Y, Maesato M, Takenaka M, Nishitsuji S, Hasegawa H, Kitagawa S (2008) Angew Chem Int Ed 47:9883–9886

18. Uemura T, Kadowaki Y, Yanai N, Kitagawa S (2009) Chem Mater 21:4096–4098
19. Alcoutlabi M, McKenna GB (2005) J Phys: Condens Matter 17:R461–R524
20. Soles CL, Ding Y (2008) Science 322:689–690
21. Uemura T, Horike S, Kitagawa K, Mizuno M, Endo K, Bracco S, Comotti A, Sozzani P, Nagaoka M, Kitagawa S (2008) J Am Chem Soc 130:6781–6788
22. Uemura T, Yanai N, Watanabe S, Tanaka H, Numaguchi R, Miyahara MT, Ohta Y, Nagaoka M, Kitagawa S (2010) Nat Commun 1:83
23. Yanai N, Uemura T, Horike S, Shimomura S, Kitagawa S (2011) Chem Commun 47:1722–1724
24. Yanai N, Kitayama K, Hijikata Y, Sato H, Matsuda R, Kubota Y, Takata M, Mizuno M, Uemura T, Kitagawa S (2011) Nature Mater 10:787–793

Chapter 6
Metal Array Fabrication Based on the Self-Organization of Metalated Amino Acids and Peptides

Hikaru Takaya

Abstract Precisely-controlled metal array fabrication can be achieved through the self-assembly of metalated amino acids and related peptides, where the functional transition-metal complexes conjugate to the α-side chains of the amino acids. The chemically and physically robust NC-half-pincer Pd- and Pt-complex-bound glutamines allowed for controlled metal array fabrication on a nanometer-scale by ultrasound-induced gelation. Control of the metal sequence was accomplished by the condensation of these Pd- and Pt-bound glutamines through the chronological ordering of the amino acid condensation processes. The supramolecular-based self-assembly of the Pd- and Pt-bound dipeptides also gave rise to nanometer-scale regulated arrays of Pd- and Pt-complexes. Furthermore NCN-pincer Pd-complex-bound norvalines were recently found to show excellent self-assembly properties, and afforded supramolecular gels together with the formation of a well-ordered Pd-array. The resulting Pd-array acted as an efficient recyclable supramolecular catalyst for the transformation of an alkynoic acid under aqueous conditions.

Keywords Amino acid • Peptide • Self-assembly • Metal array • Supramolecular gel

6.1 Background and Introduction

Bioorganometallic materials, which are the hybrids of biogenic molecules and functional organometallic compounds, have recently been recognized as fascinating materials because of their emergent properties, which are derived from their metallic and biological parts. In particular, conjugates of transition-metal

H. Takaya (✉)
Institute for Chemical Research, Kyoto University, Uji, Japan
e-mail: takaya@scl.kyoto-u.ac.jp

Y. Matsuo et al. (eds.), *Metal–Molecular Assembly for Functional Materials*,
SpringerBriefs in Molecular Science, DOI: 10.1007/978-4-431-54370-1_6,
© The Author(s) 2013

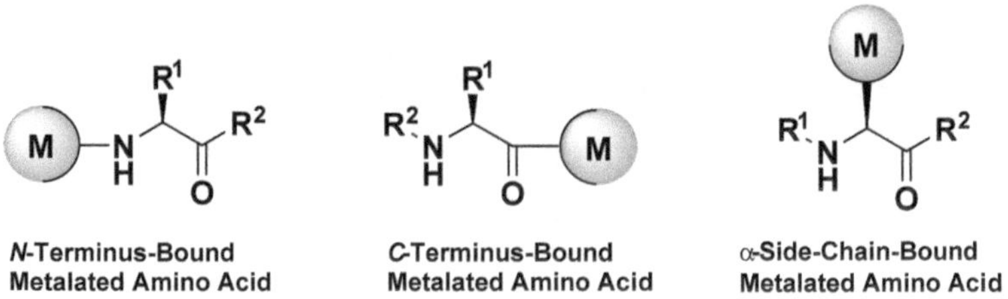

Fig. 6.1 *N*-terminus-, *C*-terminus-, and α-side-chain-bound metalated amino acids

complexes with amino acids [1, 2] and peptides [3] including proteins have been developed as functional bioorganometallic materials. The known metalated amino acids, in which the metal units are covalently bound to the amino acid scaffold, can be categorized into three different groups based on whether the metal conjugation sites are at the *N*-terminus, the *C*-terminus, or on the α-side-chain, as described in Fig. 6.1. Many different *N*- and *C*-terminus-bound metalated amino acids and their peptide derivatives have been synthesized and evaluated, mainly as medical imaging reagents. Studies on the α-side-chain-bound metalated amino acids and related peptides, however, are relatively scarce, because these compounds invariably require multistep syntheses that suffer from problems associated with racemization and metal leaching, which are less likely to occur when the metal unit is conjugated to the *N*- or *C*-terminus of the amino acids.

The first reported synthetically pure α-side-chain-metalated-bound amino acids, DL-ferrocenylalanine and DL-ferrocenylphenylalanine, were synthesized by Schlögl [4] in 1957 based on method A depicted in Scheme 6.1, where the appropriate ferrocene derivatives were transformed into the corresponding ferrocene-bound amino acid according to a multistep synthetic process. Following on from this initial report, a variety of different metallocene-bound amino acids were synthesized using method A, including some enantiomerically pure metalated amino acids. These compounds were synthesized, not only to create new bioorganometallic molecules but also for biological and medicinal applications. In a subsequent development, a one-pot method was established for the single-step installation of a metal unit (method B). This method allowed for metal ions (M) to be directly attached to a strong metal-coordination site on the α-side-chain, by taking advantage of the low coordination ability of natural functional groups such as π-coordinating aromatic groups, including phenyl alanine, tyrosine, and tryptophan and sulfur-containing functional groups, including methionine and cysteine. A variety of different artificial amino acids known as single amino acid chelates (SAACs) have been reported in the literature [5] bearing strong chelating ligands such as bipyridines, terpyridines, phenanthrolines, cyclic amines, polycarboxylic acids, bis(2-picolyl)amine and its naphthyl analogues, porphyrins, and alkynes. In the third approach towards the α-side-chain-bound metalated amino acids (method C), preformed metal complexes are conjugated to the α-side-chains of the amino acids using the appropriate coupling chemistry. This method provides a higher degree of convertibility with respect to metal selection, because it is better suited to the application of a combinatorial approach and the metal–ligand affinities do not need to be

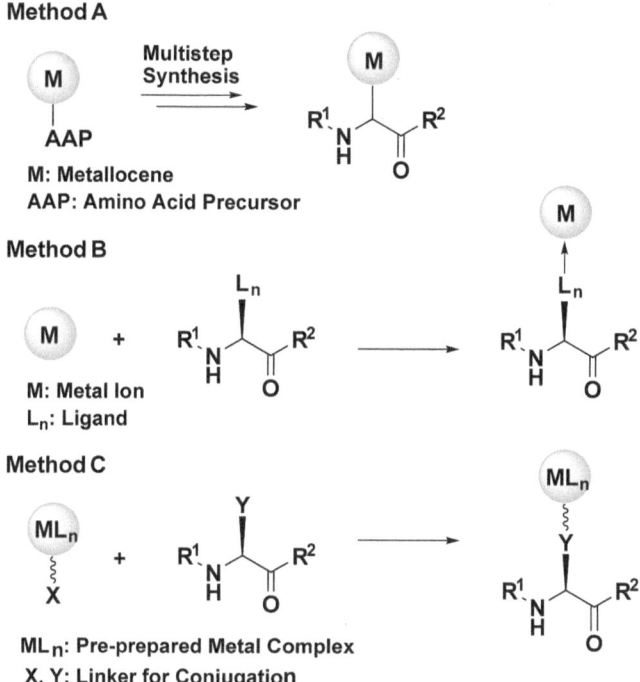

Scheme 6.1 Synthetic methods for the construction of the α-side-chain-bound metalated amino acids

evaluated using an irrational trial-and-error synthetic approach. Unfortunately, however, synthetic research based on the use of method C has been limited because of problems with the dissociation of the metal and the occurrence of racemization under basic, acidic, or high-temperature reaction conditions during the conjugation of the metal units. Since the first report of this type of synthesis by Pearson in 1986 [6], where a series of $(\eta^6$-arene)Mn(CO)$_3$-bound amino acids were synthesized by the nucleophilic aromatic substitution coupling reactions of the chlorobenzene ligand of cationic $(\eta^6$-chloroarene)Mn(CO)$_3$ complexes with tyrosine- and 4-hydroxyphenylglycine-derived phenoxide anions, several other reports for this type of synthesis have appeared in the literature. These reports, however, were focused exclusively on the development of new functional group transformation strategies for synthesis of amino acids-derived natural products rather than the development of metalated amino acids themselves. The metal units incorporated into the related amino acids were basically protecting groups for organic synthesis, and were therefore unsurprisingly labile under basic/acidic and high-temperature conditions, where they readily suffered from metal leaching. The lack of stability relating to the metal unit in these systems is practically unfavorable for their use in material research directed towards metal array-based functional materials. The first practically robust metalated amino acid prepared using method C was reported in 1996 by Jackson [7, 8], who developed a series of amino-acid-derived organozinc reagents (functionalized zinc homoenolate), where ZnI was covalently attached to the α-side-chain of a protected alanine or homoalanine. In a subsequent development, van Koten [9] reported a series of extremely robust NCN-pincer Pt- and Pd-complexes bearing 2,6-bis[(dimethylamino)methyl]benzene ligands and demonstrated that the Suzuki–Miyaura coupling of these NCN-pincer complexes with the borane adducts of protected allylglycine afforded the corresponding metalated amino acids [9]. The robust nature of these NCN-pincer complexes allowed for the preparation of the N-succinimide active ester of the NCN-pincer Pt-complex, which could be readily coupled to protected lysine residues under conventional condensation reaction conditions.

Our research group recently succeeded in synthesizing a series of NC-palladacycle-complex-bound amino acids through the conjugation of Pd- and Pt-benzaldimine complexes with the corresponding protected glutamic acids [10, 11]. The robust nature of the NC-palladacycle-complex-bound glutamic acids toward acidic/basic and high-temperature conditions enabled us to synthesize several poly-Pd and Pt-complex-bound glutamic peptides without observing any loss of the Pd- and Pt-complex moiety. These results led us to design a new type of NCN-pincer Pd-complex-bound metalated norvalines bearing a dipyridylbenzene complex of palladium PdCl(dpb) as a metal unit [12]. Metalated amino acids of this particular type showed excellent self-assembly properties and afforded well-regulated metal arrays that could be used as supramolecular catalysts [13], as well as highly emissive and electron conducting materials. Herein, the syntheses, structural determination, and application to the functional materials of the metal arrays prepared from the metalated amino acids and peptide have been reviewed in detail.

6.2 Metal Array Fabrication Though the Self-Assembly Process of Pt-Bound Glutamines

The Pt-complex-bound glutamic acid **1** was synthesized by the condensation of the N- and C-alkylated glutamic acid n-$C_{11}H_{23}$CO-Glu-NH-n-C_4H_9 with the cyclo-metalated Pt complex chloro{2-[2-hydroxyethylimino-κN]methyl]phenyl-κC}(triphenylphosphine)Pt(II) (Fig. 6.2) [10]. The cyclometalated structure of the parent NC-half-pincer Pt-complex and the stable amide bond tethering the complex and glutamine provided excellent stability towards acidic, basic, and high temperature conditions, and eliminated the possibility of the metal detaching under a variety of synthetic conditions. The self-assembly of Pt-complex-bound glutamic acid **1** efficiently proceeded in several different organic solvents to afford a supramolecular gel. Mixtures of **1** typically showed thermoreversible gelation properties (pictures a, b of Fig. 6.2) which strongly supported the notion that non-covalent interaction-induced self-assembly proceeded to give the gel. Interestingly, similar gelation behavior was also observed under ultrasound irradiation (0.45 W/cm^2 at 40.0 kHz), despite the fact that ultrasound usually destroys non-covalently bonded aggregates of peptides.

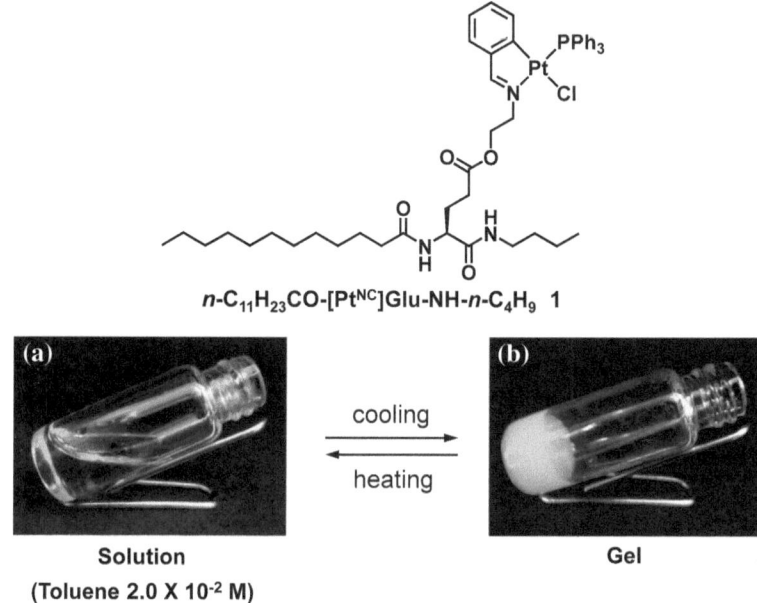

$$n\text{-}C_{11}H_{23}CO\text{-}[Pt^{NC}]Glu\text{-}NH\text{-}n\text{-}C_4H_9 \quad \mathbf{1}$$

cooling

heating

Solution

(Toluene 2.0 X 10^{-2} M)

Gel

Fig. 6.2 Self-assembly of Pt-bound glutamine

To elucidate the precise supramolecular structure of the toluene gel, the material was observed directly by scanning electron microscopy (SEM) and cryo-transmission electron microscopy (TEM). The SEM image of the supramolecular gel of **1** (Fig. 6.3a, inset) showed a bundled aggregate of micrometer-order tape-like fibrils that were typical of those normally found in amino acid and peptide-based supramolecular gels. A cryo-TEM image of a gel fibril taken under liquid helium conditions showed a fine, periodic striped structure with a spacing of approximately 2.1 nm (Fig. 6.3a). Selected area electron diffraction analysis was carried out to elucidate the self-assembled structure of **1** in the gel fibrils (Fig. 6.3b). The rectangular lattice (dashed red box) arising from the clearly orthogonal relationship of the a and c axes allowed the unit cell to be indexed under $p2gg$ two-dimensional symmetry as $a = 4.2$ and $c = 1.0$ nm.

A molecular modeling study was conducted based on these crystallographic parameters and, together with the geometrical and spectroscopic data obtained from the wide angle X-ray diffraction analysis (WAX), small angle X-ray diffraction analysis (SAX), and IR spectrum, afforded an assembly structure of **1**, as depicted in Fig. 6.4a. The 2.18 nm-interspaced lamellar structure formed by the alternating arrangement of the Pt-containing layers and alkyl chain layers matched the periodic band structure of the gel fibril in the cryo-TEM image in Fig. 6.3a. Antiparallel hydrogen-bonding association in a direction orthogonal to the ab plane was observed, as depicted in Fig. 6.4b. The 0.47 nm spacing of the staggered arrangement of **1** afforded a 1.0 nm pitch built-up along the ab plane, as indicated

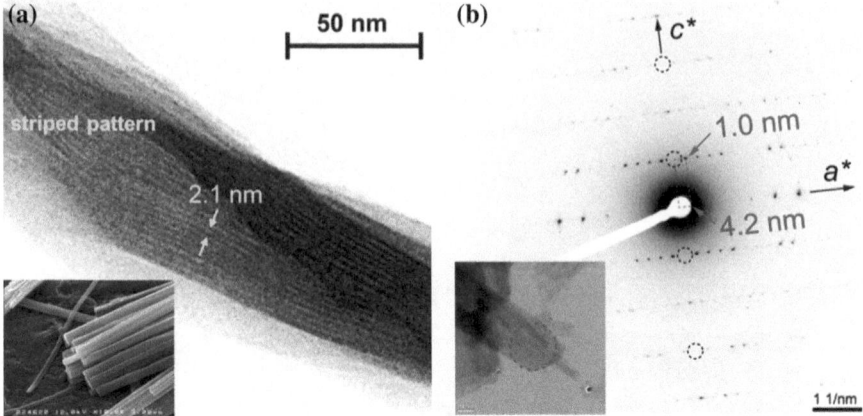

Fig. 6.3 Cryo-TEM image and selected diffractions of the fibrils of **1** in its toluene gel. The *inset* shows an SEM image and the real space image of the gel fibril

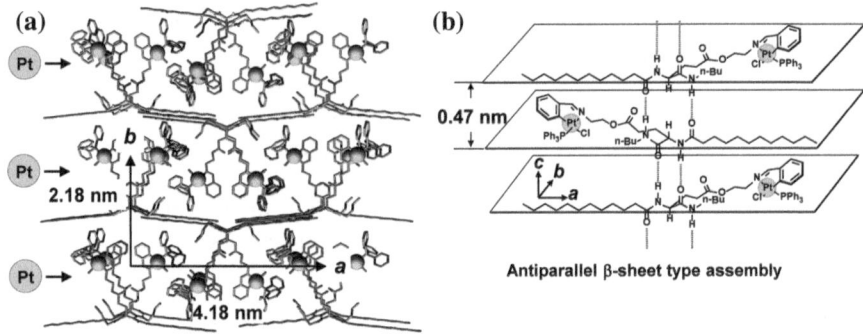

Fig. 6.4 Supramolecular structure of Pt-bound glutamine beta-sheet assembly

by the diffraction spot in Fig. 6.3b. The observed layering of the Pt complexes demonstrated that this process for the formation of supramolecular gels through the self-assembly of metalated amino acids had strong potential in the fabrication of precisely controlled metal arrays.

6.3 Sequence-Controlled Metal Array Fabrication Based on Pt- and Pd-Bound Glutamines

One of the most important features of peptide chemistry relates to the fact that sequences of amino acids can be designed and synthesized using simple condensation reaction chemistries, where the sequences are arranged by the order and number of amino acid connections. In our own work, it was envisaged that this feature could enable us to fabricate a sequence-controlled metal array on a peptide, with the metal sequence being controlled by defining the order in which the metalated amino acids were connected during the synthesis of the peptide [11, 14, 15]. The robust nature of the NC-half-pincer complexes of Pt and Pd allowed for the polymetalated peptides to be synthesized without any dissociation or scrambling of the metal. As shown in Fig. 6.5, the metal isomers of the bis-metal-conjugated dipeptides, Pd–Pd- and Pt–Pt-dipeptides, as well as the two heterometal-conjugated-dipeptides, Pd–Pt- and Pt–Pt-dipeptide (regarding the *N*-to *C*-axis orientation of the peptides) were successfully synthesized without any dissociation or scrambling of the metal.

Although the resulting heterometal sequence control of the Pd and Pt represents the minimum number of combination sets of the different metals, the possibility of metal sequence control based on the chronological order of amino acid connection was satisfactorily demonstrated. These bis-metalated peptides showed self-assembly properties under the ultrasound irradiation conditions to afford a supramolecular gel together with the formation of fibrous micro aggregates (Fig. 6.6). Structural analyses using IR, SEM, and synchrotron X-ray diffraction (XRD) data revealed that a parallel β-sheet type supramolecular assembly had formed through a series of inter-peptide amide–amide hydrogen bonding interactions. It is noteworthy that in the self-assembly of the Pd–Pt- and Pt–Pd-heterometalated dipeptides, the Pd and Pt atoms were aligned to afford a homo-metal array along the lateral sides of β-sheets (the lines of the blue or yellow balls, respectively, in Fig. 6.6). Programmable 2D-metal arrangement therefore becomes available through the direction controlled self-assembly of parallel β-sheets using heterometalated dipeptides as assembly units, which have a pre-programmed 1-D metal sequence. The mechanism of the ultrasound-induced β-sheet formation was elucidated by ^{1}H NMR association and kinetic experiments with the support of density functional theory calculation. The metalated peptides formed a self-locking structure in which the amide hydrogens and the chlorine

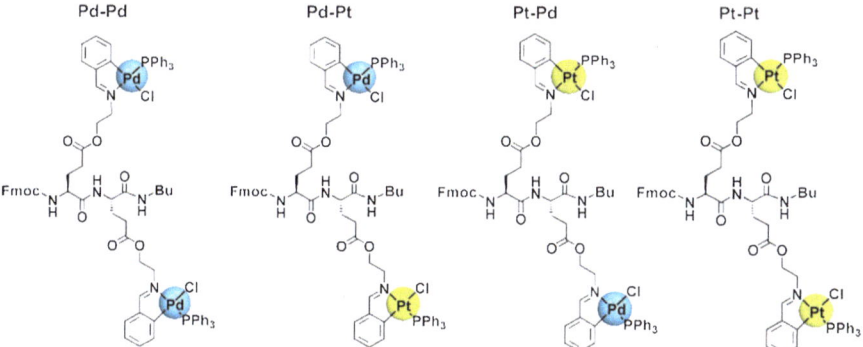

Fig. 6.5 The metal sequence-controlled dipeptides derived from the Pd- and Pt-bound glutamines

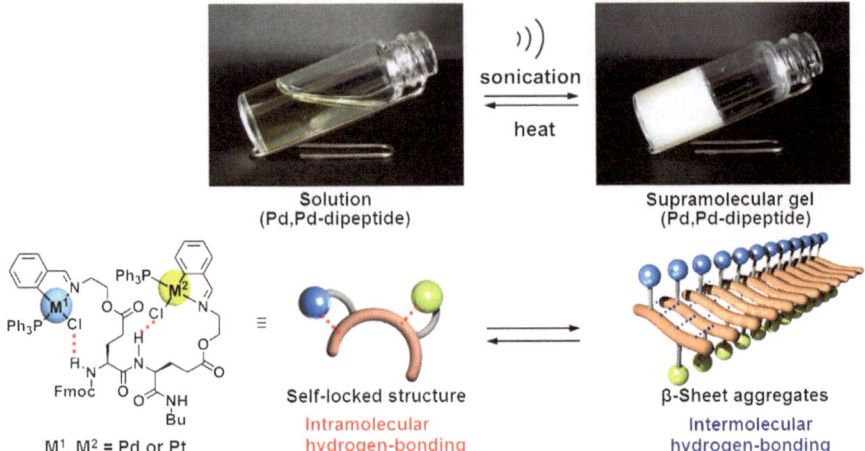

Fig. 6.6 2D-controlled metal array fabrication based on the ultrasound-induced self-assembly of Pd- and Pt-bound glutamine dipeptides

ligands of the Pd- and Pt-complexes formed N–H⋯Cl intramolecular hydrogen bond interactions. The intramolecular hydrogen bonds were subsequently cleaved and the intermolecular association of the peptide induced by the amide–amide hydrogen bonding afforded the corresponding oligomeric peptide as a nucleus for the supramolecular gel fiber. The observed first-order relationship between the gelation rate and the sonication frequencies suggested that micro cavitations bearing ultra-high thermal energies and pressures triggered the hydrogen bonding dissociation/association process.

6.4 Self-Assembled Metal Array as a Supramolecular Gel Catalyst

We recently developed a series of NCN-pincer Pd-complex-bound norvaline derivatives of the general formula P^1-D, L-[Pd^{NCN}]Nva-P^2 bearing the dipyridylbenzene palladium complex PdCl(dpb) as a metal unit (Fig. 6.7) [13, 16]. The stable bis-palladacycle structure of the PdCl(dpb) complex prevented the metal from leaching under the acidic/basic and high-temperature conditions required during the course of the conjugation process to the amino acids, as well as the subsequent deprotection and N- and C-terminus condensation reactions, including the peptide synthesis. Even though more than 60 years have passed since the first report of a metalated amino acid of ferrocenyl alanine in 1959, there have only been a few reports concerning the X-ray structure determination of these materials. Recently, we successfully elucidated the molecular structures of the NCN-pincer Pd-complex-bound norvalines using synchrotron radiation X-ray at the BL40XU

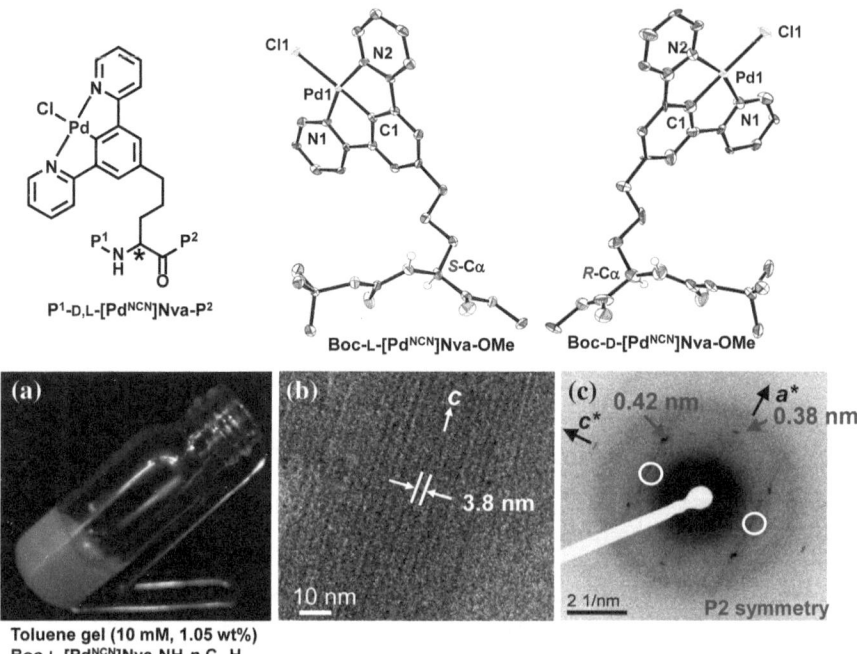

Fig. 6.7 The molecular structures of the NCN-pincer Pd-complex bound norvalines: **a** Supramolecular gel of the Pd-bound norvaline derivatives. **b** Cryo-TEM image of the toluene gel fiber. **c** Selected area diffraction image of the toluene gel fiber

beamline [17] of SPring-8. The structures of the Boc-D, L-[Pd^{NCN}]Nva-OMe were unequivocally determined with the absolute configuration at the α-carbon atom, and all of the structural parameters were found to be consistent with those of the parent Pd complex and amino acids, including the chirality of the α-carbon. This result supported the notion that the inherent physical and chemical properties of both the organometallic and amino acid parts were preserved both before and after the metal conjugation process. The self-assembly properties were evaluated based on the supramolecular gelation of *C*-long alkyl-chain substituted derivatives of Boc-L-[Pd^{NCN}]Nva-NH-*n*-C$_{11}$H$_{23}$ (Fig. 6.7a).

The well-defined metal-array formation in the gel fiber was suggested by the fine striped pattern on the gel fiber, as well as the electron diffraction pattern with *P2* symmetry obtained from the cyro-TEM analysis (Fig. 6.7b, c). The manner with which the Boc-L-[Pd^{NCN}]Nva-NH-*n*-C$_{11}$H$_{23}$ assembled in the supramolecular gel was evaluated with the support of computer modeling using the structural parameters obtained from the above cryo-TEM analysis, as well as data from the IR, WAX and SAX analyses. It ultimately afforded the well-regulated arrayed structure. As shown in Fig. 6.8, the anti-parallel β-sheet type aggregation was found to form well-regulated Pd-arrays together with amide–amide hydrogen bonding and the π–π stacking of the Pd(dpb)Cl units. These results clearly

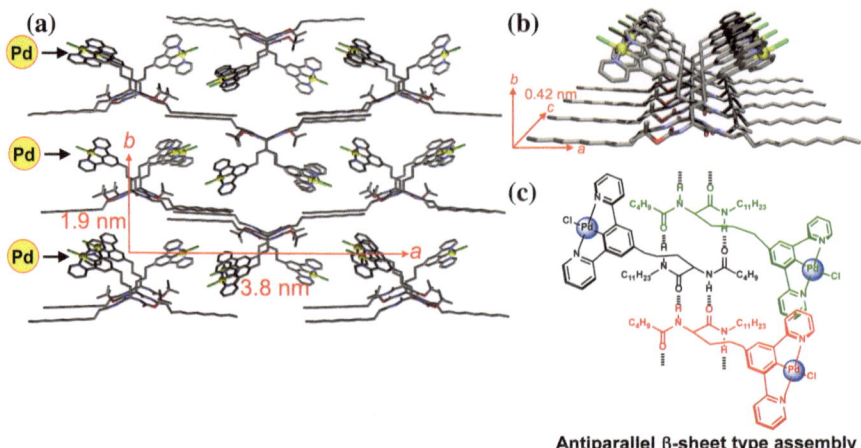

Antiparallel β-sheet type assembly

Fig. 6.8 Supramolecular structures in toluene gel of the NCN-pincer Pd-complex bound norvalines

demonstrated that the gelation process based on the self-assembly of metalated amino acids provided a highly efficient and practical method for the fabrication of a precisely controlled metal arrays that enabled us to control metal manipulation at the nanometer level according to a bottom-up approach.

Although Pd-complex based coordination polymers have been widely recognized as efficient supramolecular catalysts, supramolecular gels assembled through weak ligand–ligand interactions such as van der Waals, hydrophobic, π–π stacking, and hydrogen-bonding interactions have remained largely unexplored because of their chemical and physical fragility. We envisage that the robust nature of NCN-pincer Pd-complex-bound norvalines and their supramolecular gel will make it possible for them to be used as supramolecular gel catalysts [18]. In actual fact, the supramolecular gel of Boc-L-[Pd^{NCN}]Nva-NH-n-C$_{11}$H$_{23}$ showed efficient catalytic activity towards the cyclization of 4-alkynoic acid to afford the corresponding lactone in good yield (Fig. 6.9) [15]. It is noteworthy that the water-insoluble gel catalyst could be recovered by filtration and reused at least three times, without significant loss of catalytic activity; this catalyst afforded the cyclization product in 69 and 70 % yields when it was reused the second and the third times, respectively. In contrast, the non-gelled Pd-bound norvalines and parent PdCl(dpb) complex gave lower yields than the gel catalyst. The pictures shown in Fig. 6.9 reveal that the yellow particle of the xerogel catalyst promptly swelled through the absorption of the substrate during the reaction. This observation suggests that the substrate concentration dramatically increased the inner volume of the gel by absorption and that the reaction mainly proceeded within the inner sphere of the supramolecular gel where highly condensed Pd-catalytic sites would have formed on the large surface area of the gel to enhance the reaction. These results clearly demonstrated that the highly assembled Pd-complex array fabricated in the supramolecular gel showed unique and excellent catalytic activity and

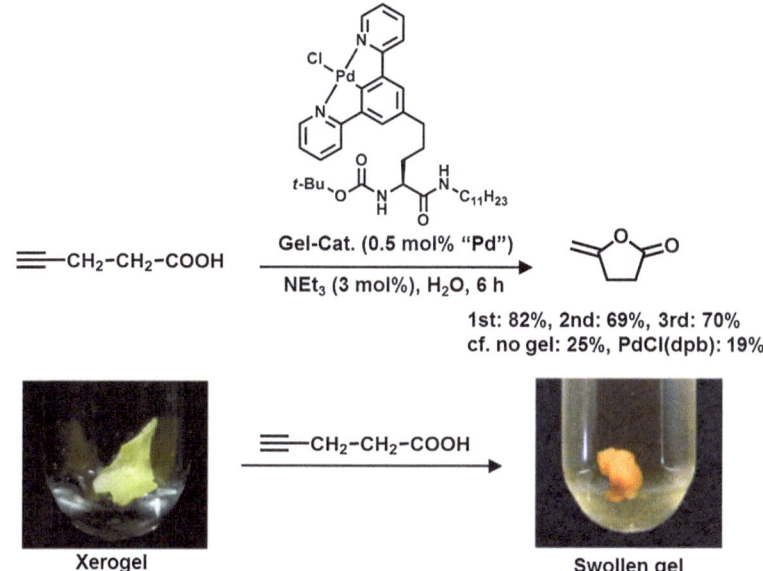

Fig. 6.9 Supramolecular gel-catalyzed cyclization

that the robust nature of the hydrogen-bonding-based supramolecular gel catalyst was firmly established by its reusability.

Acknowledgments The author would like to extend his sincerest thanks to Professor Takeshi Naota (Osaka University), Masaharu Nakamura (Kyoto University) and Katsuhiro Isozaki (Kyoto University), who were important collaborators on the work mentioned above. X-ray diffraction measurements were carried out at SPring-8 (BL40XU: 2011B1545, 2012A1161, 2012A1625, and 2012B1815).

References

1. Severin K, Bergs R, Beck W (1998) Bioorganometallic chemistry—transition metal complexes with α-amino acids and peptides. Angew Chem 110:1722–1743. doi:10.1002/19980703
2. Takaya H, Isozaki K, Nakamura M (2009) Programmable metal unit arrangement on peptides: the synthesis, structure, and self-assemble properties Pd- and Pt-bound peptides. Catal Catal 51:588–593. ISSN:0559-8958
3. van Staveren DR, Metzler-Nolte N (2004) Bioorganometallic chemistry of ferrocene. Chem Rev 104:5931–5985. doi:10.102/cr0101510
4. Schlögl K (1957) Über ferrocene-aminosäuren und verwandte verbindungen. Monatsh Chem 88:601–621. doi:10.1007/BF00901345
5. Bartholomä M, Valliant J, Maresca KP, Babich J, Zubieta J (2009) Single amino chelates (SAAC): a strategy for the design of technetium and rhenium radiopharmaceuticals. Chem Commun 493–512. doi:10.1039/B814903H

6. Pearson AJ, Bruhn PR, Hsu S-Y (1986) Preparation of diaryl ethers from tyrosine or 4-hydroxyphenylglycine using organomanganese chemistry. J Org Chem 51:2137–2139. doi: 10.1021/jo00361a043

7. Dunn, MJ, Jackson RFW, Stephenson GR (1992) Reactions of a serine-derived zinc/copper reagent with tricarbonyl(cyclohexadienyl) iron salts: synthesis of protected amino acids. Synlett 905–906. doi:10.1055/s199221536

8. Rilatt I, Caggiano L, Jackson RFW (2005) Development and applications of amino acid derived organometallics. Synlett 18:2701–2719. doi:10.1055/s2005918950

9. Guillena G, Rodríguez G, Albrecht M, van Koten G (2002) Covalently bonded platinum(II) complexes of α-amino acids and peptides as a potential tool for protein labeling. Chem Eur J 8:5368–5376. doi:10.1002/15213765

10. Isozaki K, Ogata K, Haga Y, Sasano D, Ogawa T, Kurata H, Nakamura M, Naota T, Takaya H (2012) Metal array fabrication through self-assembly of Pt-complex-bound amino acids. Chem Commun 48:3936–3938. doi:10.1039/c2cc117530d

11. Isozaki K, Takaya H, Naota T (2007) Ultrasound-induced gelation of organic fluids with metalated peptides. Angew Chem Int Ed 46:2855–2857. doi:10.1002/anie.200605067

12. Ogata K, Sasano D, Yokoi T, Isozaki K, Seike H, Yasuda N, Ogawa T, Kurata H, Takaya H, Nakamura M (2012) Synthesis and supramolecular association of NCN-pincer Pd-complex-bound norvaline derivatives toward fabrication of controlled metal array. Chem Lett 41:194–196. doi:10.1246/cl.2012.194

13. Ogata K, Sasano D, Yokoi T, Isozaki K, Seike H, Takaya H, Nakamura M (2012) Pd-complex-bound amino acid-based supramolecular gel catalyst for intramolecular addition-cyclization of alkynoic acids in water. Chem Lett 41:498–500. doi:10.1246/cl.2012.498

14. Isozaki K, Haga Y, Ogata K, Naota T, Takaya H (2013) Metal array fabrication based on Ultrasound-induced self-assembly of metalated dipeptides. Dalton Trans doi:0.1039/C3DT51696B

15. Takahashi E, Takaya H, Naota T (2010) Dynamic vapochromic behaviors of organic crystals on the open-close motions of S-shaped donor-acceptor folding units. Chem Eur J 16:4793–4802. doi:10.1002/chem.200903403

16. Ogata K, Sasano D, Yokoi T, Isozaki K, Yoshida R, Takenaka T, Seike H, Ogawa T, Kurata H, Yasuda N, Takaya H, Nakamura M (2013) NCN-pincer pd-complex-bound norvalines: Synthesis and self-assembly. Chem Eur J doi:10.1002/chem.201301513

17. Yasuda N, Murayama H, Fukuyama Y, Kim J, Kimura S, Toriumi K, Tanaka Y, Moritomo Y, Kuroiwa Y, Kato K, Tanaka H, Takata M (2009) X-ray diffractometry for the structure determination of a submicrometre single powder grain. J Synchrotron Rad 16:352–357. doi:10.1107/S090904950900675X

18. Takaya H, Iwaya T, Ogata K, Isozaki K, Yokoi T, Yoshida R, Yasuda N, Seike H, Takenaka T, Nakamura M (2013) Synthesis, structure, and function of PCP pincer transition-metal-complex-bound norvaline derivatives PCP-Pincer transition-metal-complex-bound norvaline derivatives synlett accepted

Chapter 7
Coordination Chemistry in Self-Assembly Proteins

Takafumi Ueno

Abstract Bioinorganic chemistry represents an area of growing importance in the development of nanomaterials and sustainable energy sources, because the coordination of metals in biological systems can effectively promote elaborate enzymatic reactions, such as photosynthesis, nitrogen fixation and biomineralization. Although such systems employ protein assemblies as molecular scaffolds, the important roles of protein assemblies have not yet been systematically investigated. We have recently published a number of reports concerning the rational design of protein assemblies for the integration of catalytic reactions with metal complexes, as well as the preparation of biominerals and mechanistic investigations of biomineralization processes with protein assemblies.

Keywords Bioinorganic chemistry • Metalloprotein • Self-assembly protein • Coordination chemistry • Protein cage

7.1 Molecular Scaffolds for Metal Complexes

Proteins can serve as useful platforms for the construction of metal-molecular assemblies. Many essential enzymes, such as cytochrome P450, nitrogenase, and photosystem II, contain metal cofactors in their active sites. The properties of the cofactors are regulated within the unique environments of the proteins [1]. On the basis of molecular design within natural systems, various metal-protein hybrids have been prepared according to a variety of different techniques, including (1) the covalent conjugation of metal complexes to proteins; (2) the alternation of amino acids around the active sites

T. Ueno (✉)
Graduate School of Bioscience and Biotechnology, Tokyo Institute of Technology,
Yokohama, Japan
e-mail: tueno@bio.titech.ac.jp

Y. Matsuo et al. (eds.), *Metal–Molecular Assembly for Functional Materials*,
SpringerBriefs in Molecular Science, DOI: 10.1007/978-4-431-54370-1_7,
© The Author(s) 2013

of known enzymes; and (3) the modification of metal cofactors or substrates [2, 3]. Unfortunately, however, these methods invariably involve complex procedures, which have substantially limited the expansion of this field. If synthetic metal cofactors could be readily accommodated within protein scaffolds, they could be applied to the preparation of a variety of different artificial metal-protein hybrids. We recently developed a technique for the non-covalent insertion of metal complexes into proteins [4, 5], and this technique represents one of the simplest methods currently available for preparation of metalloproteins. In this chapter, recent progress towards the development of techniques for the preparation of metal-protein assemblies will be discussed.

7.2 Construction of Artificial Metalloprotein

Natural metal cofactors are biologically functionalized to promote a variety of different catalytic reactions with high levels of reactivity and selectivity when they are fixed in or onto proteins. Based on our understanding of the natural systems, it is generally believed that the functions of synthetic metal complexes can be controlled within the active sites of proteins by an alternative fixation for native metal cofactors. With this in mind, we constructed a series of apomyoglobin (apo-Mb) and Schiff base complex composites to test this hypothesis [6]. Mb, which is a hemeprotein responsible for O_2 storage in muscle, has a cavity with a diameter of approximately 10 Å to accommodate heme (Fig. 7.1a) [1]. The affinity of heme for apo-Mb results from a number of factors, including (1) hydrogen bonding and hydrophobic interactions between the protein and the heme; and (2) the coordination of the proximal His93 to the heme iron [7]. Thus, it should be possible to design Schiff base complexes suitable for incorporation in Mb on the basis of these considerations.

There are several advantages to using Schiff base complexes as substitutes for heme, because the molecular size and coordination geometry of Schiff base complexes are almost identical to those of heme, and size and hydrophobic properties of the Schiff base ligands can be readily manipulated. A typical reconstitution procedure for a Schiff base complex with apo-Mb is shown in Fig. 7.1a. Apo-Mb was obtained according to the HCl/2-butanone method previously reported in the literature [8], and a methanol solution of a Schiff base complex was then added into the apo-Mb solution to obtain a Schiff base complex/Mb composite [6]. The relative stability of the composite was estimated by electrospray ionization time of flight mass spectroscopic analysis of the reaction mixture. The results indicated that 3,3′-Me$_2$-salophen (**1**) was the most stable ligand fixed in the active site of apo-Mb among the ligands screened [4]. The crystal structure of the **Fe·1·apo-Mb** composite was determined but the exact position of the complex in the active site could not be determined because the weak electron density of the complex resulted in a high level of flexibility and multiple conformations of the complex at the active site (Fig. 7.1b) [9].

Computer modeling was conducted based on the crystal structure and indicated that the single mutation of the Ala residue at position 71 to a Gly residue

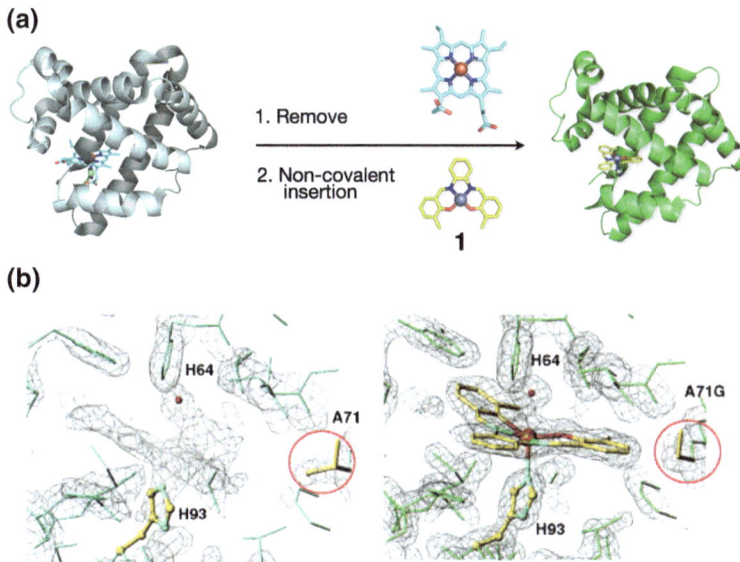

Fig. 7.1 **a** Schematic drawing of the non-covalent insertion of a synthetic metal complex into apo-Mb. **b** (*left* and *right*) Crystal structures of the active centers of **Fe·1·apo-Mb** and **Fe·1·apo-A71GMb**, respectively. Reproduced with permission from the American Chemical Society, see Ref. [10]

would effectively fix the complex. The crystal structure of **Fe·1·apo-A71GMb** clearly shows the intact electron density of the complex existing in the active site (Fig. 7.1b). Moreover, the binding rate of CN^- of **Fe·1·apo-A71GMb** was 250 times faster than that of **Fe·1·apo-Mb**. This value was very similar to that of the original holo-Mb. The apo-Mb cavity is not only capable of accommodating Schiff base complexes for catalyzing oxygenation reactions but is also capable of accommodating organometallic complexes with different coordination geometries [6, 10–12].

An artificial protein–protein electron transfer (ET) complex was constructed by employing heme oxygenase (HO), which catalyzes the conversion of heme to biliverdin, in conjunction with NADPH-cytochrome P450 reductase (CPR) [13]. Crystal structures of the HOs revealed the presence of an invariant hydrogen bond between the Arg177 residue and the O atom of the heme propionate-7 (Fig. 7.2a) [14–16]. The mutagenesis experiments concluded that this residue was an important residue for the formation of a transient ET complex with CPR [17]. Several attempts were made to incorporate an Fe^{III}(Schiff base), which could form the appropriate hydrogen bonding interaction with the HO, at the active site, and accelerate the reduction process involving the NADPH-CPR (Scheme 7.1) [18]. For the **Fe·2·HO**, the **Fe·2** was found to have the same hydrogen bond to Arg177 (Fig. 7.2b) as that observed for the heme·HO complex, indicating that **Fe·2** was capable of receiving an electron from the CPR. In contrast, the Arg177 residue

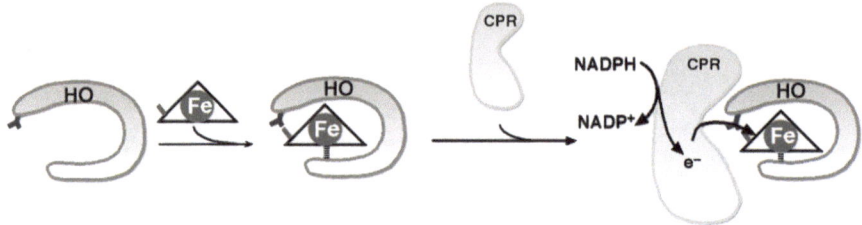

Scheme 7.1 Schematic representation of the electron-transfer reaction between the Schiff base complex in the HO/CPR complex and NADPH

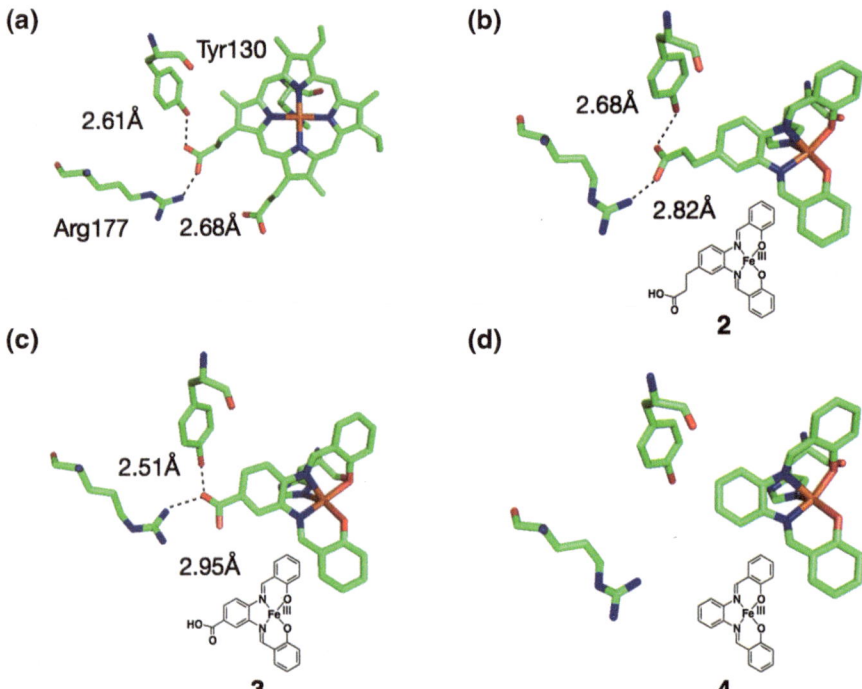

Fig. 7.2 **a** The structure of the HO active site reconstructed with heme, **b** Fe·**2**, **c** Fe·**3**, **d** Fe·**4**

in **Fe·3·HO** was not positioned in such a way that it could assist the ET because one oxygen atom of the carboxyl group in the **Fe·3** formed two hydrogen bonding interactions with Nη¹ (Arg177) and Oη (Tyr130) (Fig. 7.2c).

Furthermore, **Fe·4** did not have a hydrogen bond with Arg177 (Fig. 7.2d). On the basis of these results, it was envisaged that the Arg177 residue in the **Fe·3·HO** and **Fe·4·HO** complexes would not have a pronounced effect on the ET reaction with CPR. The reduction rate of the composites with NADPH-CPR was dependent on their hydrogen bonding interactions (i.e., **Fe·2** > **Fe·3** > **Fe·4**), whereas the reduction rates of the composites with $Na_2S_2O_4$ were effectively dependent only on their redox potentials (i.e., **Fe·4** > **Fe·3** > **Fe·2**). These results suggested that the hydrogen bonding network provided by the Arg177 residue and the propionate in **Fe·2** directly participated in the ET reaction of the HO/NADPH-CPR system. These results therefore represent the first example of a synthetic metal complex activated through a protein–protein ET system, and demonstrate that the hydrogen bond is crucial to the ET reactions of metalloproteins [19, 20].

7.3 Accumulation of Metals in a Protein Assembly Cage of Ferritin

Molecular assemblies of protein cages have been used for the introduction of coordination compounds to the interior and exterior surfaces of proteins because they are thermally stable and the cages can be readily modified both chemically and genetically [21]. Ferritin (Fr) is composed of 24 subunits and could be used as an iron storage cage with an internal diameter of 8 nm. Threefold axis channels are formed at the positions where the three adjacent subunits intersect. These channels provide a pathway allowing for the penetration of foreign molecules or metal ions into the protein cavity [22]. Although the subunits are assembled by non-covalent interactions, Fr is stable enough to retain an intact cage structure at temperatures up to 80 °C and pH values in the range of 2–11 [23]. These properties have allowed for a variety of different nanoparticles of Au, Ag, Pd, CdS and CdSe to be synthesized within the apo-Fr cage [24–28]. Moreover, apo-Fr has been used as a vehicle for a large number of diverse metal complexes [29–33]. We have developed a series of catalytic reactions within the interior space of Fr using metal compounds.

The properties of Fr for the accumulation of metal ions as well as the deposition of metal nanoparticles were used to prepare a Pd nanoparticle capable of olefin-hydrogenation [25]. When Pd(II) ions were reacted with apo-Fr in the reaction mixture, Pd(II) ions were spontaneously captured in the cage (Fig. 7.3a). Following the addition of $NaBH_4$ as a reducing agent in a buffered solution, the ions were reduced to form a single nanoparticle in the apo-Fr with a well monodispersed size distribution. The composite could then catalyze the olefin-hydrogenation of acryl amide and its derivatives with a higher level of reactivity than that of the naked Pd nanoparticles under the same conditions (Fig. 7.3b). Given that the substrates should be able to penetrate through the three-fold axis channel, the catalytic reactions occurring within the inside exhibit size-selectivity. The reactivity of the composite was improved via the formation of a Pd-Au bimetallic nanoparticle [34]. Thus, a gold nanoparticle was initially prepared in the apo-Fr cage, Pd ions were then sequentially accumulated and reduced to form shell structure on the preformed Au nanoparticle. The core(Au)-shell(Pd) type bimetallic nanoparticle was

Fig. 7.3 Schematic
representation of (**a**) the
preparation of **Pd·apo-Fr**,
(**b**) olefin hydrogenation with
Pd·apo-Fr

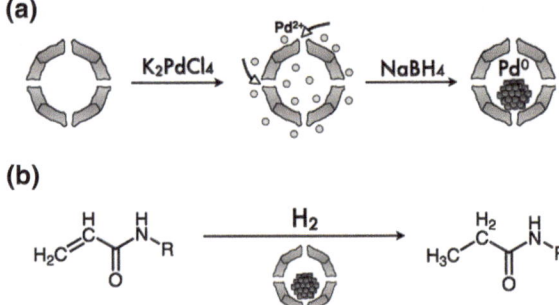

about four times more reactive than the original pure Pd nanoparticle in apo-Fr. This hybrid assembly of the protein and the nanoparticle could serve as a nanoreactor for promoting heterogeneous catalytic reactions.

Attempts were then made to incorporate metal complex catalysts within the apo-Fr (Fig. 7.4) [29]. When Rh(2,5-norbornadiene = nbd) complexes were reacted with apo-Fr, the complexes were quantitatively accumulated on specific binding sites on the interior surface of apo-Fr, as confirmed with inductively coupled plasma and X-ray structure analyses. The composite was capable of catalyzing the polymerization reaction of phenylacetylene and its derivatives in the cage with restricted molecular weights and a well monodispersed size distribution. In general, when phenylacetylene monomers are polymerized with Rh(nbd) in aqueous solution, the oligomers are precipitated with a polydispersed size distribution because of the relatively reduced level of solubility of the monomer and oligomer units in the aqueous media. The discrete space associated with the molecular assembly serves as a nanoreactor that is capable of controlling the number of monomer units accumulated within the reactor and allows the catalytic reactions to proceed smoothly and efficiently. The properties of this apo-Fr system could be applied to the reactions of a variety of different metal complexes. When Pd(allyl) complexes were fixed within the apo-Fr cage, they could be used to catalyze Suzuki–Miyaura coupling reactions in apo-Fr [30]. Ferrocenes deposited in the apo-Fr cage were used to control the electrochemical properties of the complex [35].

Fig. 7.4 Incorporation
of Rh(nbd) and the
polymerization reaction.
Reproduced with permission
from the American Chemical
Society, see Ref. [29]

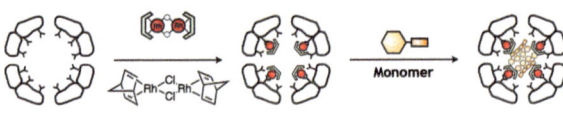

7.4 Conclusions and Perspectives

In this chapter, we have discussed the potential for proteins to accommodate a variety of different metal ions, metal nanoparticles, and metal complexes. The functions of the resulting composites can be controlled using recently developed protein engineering techniques. There are, however, still some important challenges to face in terms of designing the multiple functions required of these materials and improving the stability of these proteins to resemble those of the proteins in natural systems. In nature, self-assembling proteins can serve as molecular platforms capable of promoting multiple enzymatic reactions with many metal cofactors precisely arranged in the molecular space. We believe that studies of these native systems will provide intriguing information for the future design of protein assemblies with multiple functions.

References

1. Bertini I, Gray HB, Stiefel EI, Valentine JS (2007) Biological inorganic chemistry—structure and reactivity. University Science Books, Sausalito
2. Qi DF, Tann CM, Haring D, Distefano MD (2001) Chem Rev 101:3081–3111
3. Lu Y, Berry SM, Pfister TD (2001) Chem Rev 101:3047–3080
4. Ueno T, Koshiyama T, Abe S, Yokoi N, Ohashi M, Nakajima H, Watanabe Y (2007) J Organomet Chem 692:142–147
5. Ueno T, Abe S, Yokoi N, Watanabe Y (2007) Coord Chem Rev 251:2717–2731
6. Ohashi M, Koshiyama T, Ueno T, Yanase M, Fujii H, Watanabe Y (2003) Angew Chem Int Ed 42:1005–1008
7. Hunter CL, Lloyd E, Eltis LD, Rafferty SP, Lee H, Smith M, Mauk AG (1997) Biochemistry 36:1010–1017
8. Ascoli F, Fanelli M, Antonini E (1981) Methods Enzymol 76:72–87
9. Ueno T, Ohashi M, Kono M, Kondo K, Suzuki A, Yamane T, Watanabe Y (2004) Inorg Chem 43:2852–2858
10. Abe S, Ueno T, Reddy PAN, Okazaki S, Hikage T, Suzuki A, Yamane T, Nakajima H, Watanabe Y (2007) Inorg Chem 46:5137–5139
11. Satake Y, Abe S, Okazaki S, Ban N, Hikage T, Ueno T, Nakajima H, Suzuki A, Yamane T, Nishiyama H, Watanabe Y (2007) Organometallics 26:4904–4908
12. Ueno T, Koshiyama T, Ohashi M, Kondo K, Kono M, Suzuki A, Yamane T, Watanabe Y (2005) J Am Chem Soc 127:6556–6562
13. Chu GC, Katakura K, Zhang XH, Yoshida T, Ikeda-Saito M (1999) J Biol Chem 274:21319–21325
14. Schuller DJ, Wilks A, de Montellano PRO, Poulos TL (1999) Nat Struct Biol 6:860–867
15. Sugishima M, Omata Y, Kakuta Y, Sakamoto H, Noguchi M, Fukuyama K (2000) Febs Lett 471:61–66
16. Hirotsu S, Chu GC, Unno M, Lee DS, Yoshida T, Park SY, Shiro Y, Ikeda-Saito M (2004) J Biol Chem 279:11937–11947
17. Wang JL, de Montellano PRO (2003) J Biol Chem 278:20069–20076
18. Ueno T, Yokoi N, Unno M, Matsui T, Tokita Y, Yamada M, Ikeda-Saito M, Nakajima H, Watanabe Y (2006) Proc Nat Acad Sci USA 103:9416–9421
19. Wuttke DS, Bjerrum MJ, Winkler JR, Gray HB (1992) Science 256:1007–1009
20. Babini E, Bertini I, Borsari M, Capozzi F, Luchinat C, Zhang XY, Moura GLC, Kurnikov IV, Beratan DN, Ponce A, Di Bilio AJ, Winkler JR, Gray HB (2000) J Am Chem Soc 122:4532–4533

21. Uchida M, Klem MT, Allen M, Suci P, Flenniken M, Gillitzer E, Varpness Z, Liepold LO, Young M, Douglas T (2007) Adv Mater 19:1025–1042
22. Theil EC (1987) Annu Rev Biochem 56:289–315
23. Santambrogio P, Levi S, Arosio P, Palagi L, Vecchio G, Lawson DM, Yewdall SJ, Artymiuk PJ, Harrison PM, Jappelli R, Cesareni G (1992) J Biol Chem 267:14077–14083
24. Butts CA, Swift J, Kang S-g, Di Costanzo L, Christianson DW, Saven JG, Dmochowski IJ (2008) Biochemistry 47:12729–12739
25. Ueno T, Suzuki M, Goto T, Matsumoto T, Nagayama K, Watanabe Y (2004) Angew Chem Int Ed 43:2527–2530
26. Wong KKW, Mann S (1996) Adv Mater 8:928–932
27. Yamashita I, Hayashi J, Hara M (2004) Chem Lett 33:1158–1159
28. Zhang L, Swift J, Butts CA, Yerubandi V, Dmochowski IJ (2007) J Inorg Biochem 101:1719–1729
29. Abe S, Hirata K, Ueno T, Morino K, Shimizu N, Yamamoto M, Takata M, Yashima E, Watanabe Y (2009) J Am Chem Soc 131:6958–6960
30. Abe S, Niemeyer J, Abe M, Takezawa Y, Ueno T, Hikage T, Erker G, Watanabe Y (2008) J Am Chem Soc 130:10512–10514
31. Aime S, Frullano L, Crich SG (2002) Angew Chem Int Ed 41:1017–1019
32. Lucon J, Abedin MJ, Uchida M, Liepold L, Jolley CC, Young M, Douglas T (2010) Chem Commun 46:264–266
33. Yang Z, Wang X, Diao H, Zhang J, Li H, Sun H, Guo Z (2007) Chem Commun 3453–3455
34. Suzuki M, Abe M, Ueno T, Abe S, Goto T, Toda Y, Akita T, Yamadae Y, Watanabe Y (2009) Chem Commun 4871–4873
35. Niemeyer J, Abe S, Hikage H, Ueno T, Erker G, Watanabe Y (2008) Chem Commun 6519–6521

Chapter 8
Two-Dimensional Alignment of Conjugated Polymers

Masayuki Takeuchi

Abstract Orientated polymers and polymer nanostructures have attracted considerable attention because of their numerous potential applications. As well as reporting several methods for the organization of conjugated polymers, we have also reported a series of novel concepts for the alignment of conjugated polymers according to the action of supramolecular bundling molecules. In this chapter, two different approaches for aligning conjugated polymers will be introduced, including (1) the use of a cross-linking molecule ('aligner'); and (2) the use of a twining polymer ('twimer') for the organization of conjugated polymers. Aligner molecules effectively bind and crosslink conjugated polymers in a positive allosteric manner to form organized supramolecular assemblies in solution. The cast films resulting from these solutions can result in a crystalline sheet with periodicities corresponding to the distances between the polymers, as confirmed by several microscopic studies. In contrast, twimers wrap themselves around the conjugated polymers in a helical manner to form one-dimensional complexes in solution, and these complexes subsequently self-assemble into two-dimensionally aligned structures during the drying process. The resulting films of conjugated polymer–twimer composites provide highly ordered, crystalline structures. The principal results obtained in the solution and solid states are discussed herein.

Keywords Conjugated polymers • Supramolecular chemistry • Molecular recognition • Alignment

M. Takeuchi (✉)
Organic Materials Group, National Institute for Materials Science, Tsukuba, Japan
e-mail: TAKEUCHI.Masayuki@nims.go.jp

Y. Matsuo et al. (eds.), *Metal–Molecular Assembly for Functional Materials*,
SpringerBriefs in Molecular Science, DOI: 10.1007/978-4-431-54370-1_8,
© The Author(s) 2013

8.1 Introduction

π-Conjugated molecules have been the subject of considerable levels of attention because of their potential applications in molecular devices as an alternative to inorganic materials. The development of novel approaches enabling control over the nanostructures of π-conjugated oligomers and polymers has become an increasingly important area of research in materials science, which has led to the discovery of a variety of different functional materials with optimized π-electronic properties [1–4]. Nanostructures constructed from conjugated oligomer/polymer blends, in particular, have received growing levels of interest on account of their numerous applications, which include electronic devices such as field-effect transistors, light-emitting diodes, and photovoltaic cells [5, 6]. Supramolecular approaches have also been applied to the development of techniques for the organization of conjugated oligomers and polymers in a non-covalent manner, including the use of host matrices, liquid crystalline phases, and air–water interfaces [7–9]. From a supramolecular perspective, we recently proposed a series of new approaches towards the alignment of conjugated polymers to form two-dimensional (2-D) assemblies. In this chapter, supramolecular cross-linking molecules ('aligners') and twining polymers ('twimers') will be introduced. These materials possess conjugated polymer recognition sites and can align conjugated polymers in a non-covalent manner. The concepts introduced in this chapter will complement the existing techniques already in use for the supramolecular and macromolecular assembly of 1-D conjugated polymers.

8.2 Design of Synthetic Cross-Linkers (Aligners)

The aligner molecules **1** and **2** were designed bearing co-facial porphyrinatozinc moieties that would enable them to sandwich and bundle diamino-functionalized conjugated polymers (CPs) within their clefts through high affinity coordination bonds (Fig. 8.1) [10]. Although the distances between the pairs of porphyrinatozinc

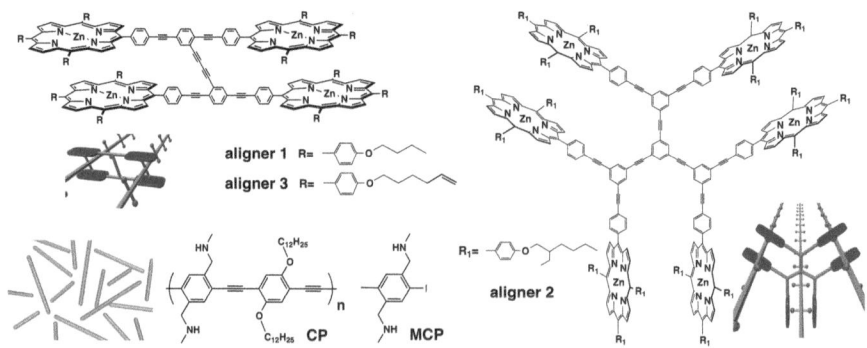

Fig. 8.1 Structures of aligners **1–3**, and CP

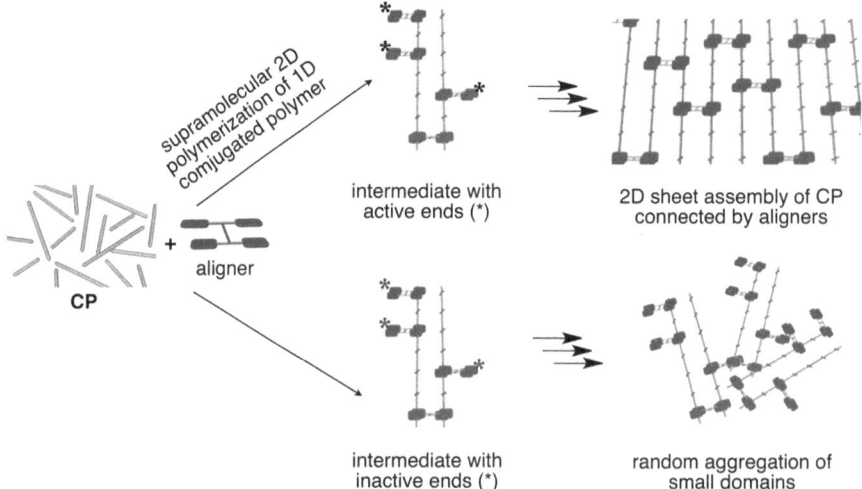

Fig. 8.2 Schematic representation of how positive allosterism facilitates the formation of the supramolecular assembly. (*top*) In a positive allosteric system, the first polymer binding event to the porphyrinatozinc cleft predisposes the second binding site (denoted by the *red asterisk*), resulting in an even higher level of affinity towards the second polymer . The CPs were expected to form aligned 2-D assemblies (*bottom*). Without allosterism, the first polymer-binding event does not enhance the second binding affinity (denoted by the *blue asterisk*), leading to formation of a random aggregation of small domains, instead of the large aligned assemblies

moieties aligned in a parallel manner in **1** and **2** varied as a consequence of conformational rearrangements about their rotational axes (i.e., the butadiyne unit of **1** and the ethynylene bridges of **2**), the distances between the clefts when in a co-facial orientation were 2.5 and 2.0 nm, respectively. Furthermore, each pair of co-facially aligned porphyrinatozinc moieties in **1** and **2** could bind to a diamine moiety belonging to the polymer in a positive allosteric manner [11] to organize the polymers into aligned, rather than random, assemblies. Thus, in this process, the binding of the first polymer to an aligner molecule effectively facilitated the second binding event, which resulted in the facile formation of the aligned assemblies (Fig. 8.2).

8.3 Two Dimensional Alignment of Poly(Phenylene Ethynylene)

The complexation behaviors between the aligners and the CPs in solution were observed by UV-Vis spectroscopy. To confirm the cooperative nature of the binding processes between the CPs and aligner **1** and **2**, 1,4-di-*N*-metyhylaminomethybenzene (MCP), which is the monomer analog of the CP, was used as guest molecule for the aligners. The formation of the [1 · MCP] or [2 · MCP] complex in CHCl3 was observed on the basis of the bathochromic shifts in the Soret and Q bands of aligners 1 and 2 in the UV-vis absorption spectra following the successive addition of MCP. The cooperative nature of the MCP-binding process was confirmed using a nonlinear least-squares method and Hill plot, assuming a multi-step binding mechanism. As expected, the addition of the CPs to a solution of 1 or 2 in CHCl3 resulted in bathochromic shifts in the Soret and Q bands similar to those observed for 1 or 2 following the addition of MCP. The complexation and bundling behaviors between the CPs and aligners were also supported by fluorescence spectroscopic studies. The emission intensity of the CPs in CHCl3 decreased, without any change in its shape, when it was mixed with the aligners. This phenomenon likely occurred as a consequence of the efficient energy transfer from the CPs to 1 within the complex, because the emission wavelength of CP overlapped considerably with the absorption band of 1.

To obtain a greater insight into the morphological properties of the CP · aligner complexes, we investigated whether the supramolecular polymer bundles were maintained in the solid state by means of atomic force microscopy (AFM) and transmission electron microscopy (TEM). An AFM image of the CPs mixed with aligner 1 revealed that the assemblies were well dispersed in a rectangular shape on the highly ordered pyrolytic graphite. Although the observed area of the overall [1 · CP] assembly was 20- to 25-fold larger in homothetic shape, the height was remained largely unchanged relative to that observed for the CP itself. On the basis of this 2-D polymerization of CP, it was envisaged that 1 must bundle with CP to form the aligned supramolecular assemblies with the observed solid-state morphology. High resolution (HR)-TEM images of the assembly of CP and 1 revealed that the materials existed as a crystalline sheet, which featured a multi-lamellar morphology with a periodicity of 2.0 nm over a distance of 200 nm. The periodicity of 2.0 nm corresponded to the distance between the CP units when they were bundled in a parallel manner. From these AFM and TEM analyses, it was deduced that the aligner assembled the CPs parallel to each other in its clefts to form the highly ordered structures that could be affected by both the geometry of the aligner and the chemical structures of the CPs.

When olefinic moieties were introduced at the peripheral positions of aligner 1, the resulting aligner 3 [12, 13] was converted into the corresponding bicyclic structure by ring-closing olefin metathesis (RCM). It was envisaged that the RCM reaction would proceed more efficiently following the complexation with the CPs as templates. Furthermore, the conversion to the poly-pseudo-rotaxane structures should stabilize the CP assemblies. Indeed, the RCM reaction of the supramolecular [aligner 3 · CP] assemblies successfully proceeded (Fig. 8.3). The resultant poly-pseudo-rotaxane structures were sufficiently stable to permit the separation of the assemblies of the CPs by size exclusion chromatography. Micro-meter sized crystalline sheets were observed in the HR-TEM images of the immobilized [3 · CP] complex. A multilamellar morphology with a periodicity of 2.0 nm was observed, which was identical to the value obtained for the [1 · CP] assembly.

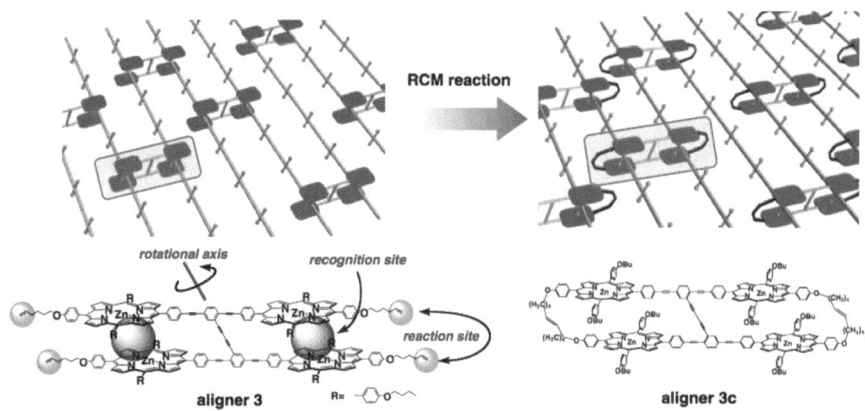

Fig. 8.3 Schematic illustration for the alignment of the conjugated polymer CP mediated by **3** and the conversion to the poly-pseudo-rotaxane structures **3c**

8.4 Alignment of Polyanilines and Alternating Arrays of Different Conjugated Polymers

This supramolecular cross-linking strategy for the production of 2-D aligned assembly of CPs can be applied to polyaniline (i.e., emeraldine base (EB) and emeraldine salt (ES)) [14]. A palladium complex was selected to design an aligner molecule for cross-linking EB and ES, because it was envisaged that this complex would show a high affinity towards the imine moiety in EB. Aligner molecule **4** was subsequently designed bearing two palladium tweezers, with each of these tweezers consisting of two palladium complexes (Fig. 8.4). The palladium complex-based binding sites in aligner **4** could rotate freely around the phenylene-1,4-diyne axis, whereas the rotation was suppressed when the binding of the guest polymer (polyanilines) to the clefts occurred. As a result, each pair of the palladium complexes in aligner **4** were cross-linked to the diimine moieties of EB and ES in an allosteric manner to form polymer bundles. In these polymer bundles, the distance between the two binding subunits was estimated to be approximately 2.3 nm using computational methods. The polyaniline-bundling process was readily monitored and analyzed by UV-vis spectroscopy. The solution-cast films of the **4** · EB assemblies were then prepared on a TEM grid, and micron-sized sheet morphologies with clear electron diffraction patterns were observed under the conditions of [**4**]/[EB]unit = 0.20:1 to 0.25:1. TEM images of the **4** · EB assemblies ([**1**]/[EB]unit = 0.20:1) revealed that the periodicity of the multilamellar contrast was 2.5 nm over a distance of a few hundred nanometers, as determined from the Fourier-filtered image. It is noteworthy that the distance of 2.5 nm was almost identical to the distance calculated when the EBs were bundled by **4** in an anti-conformation.

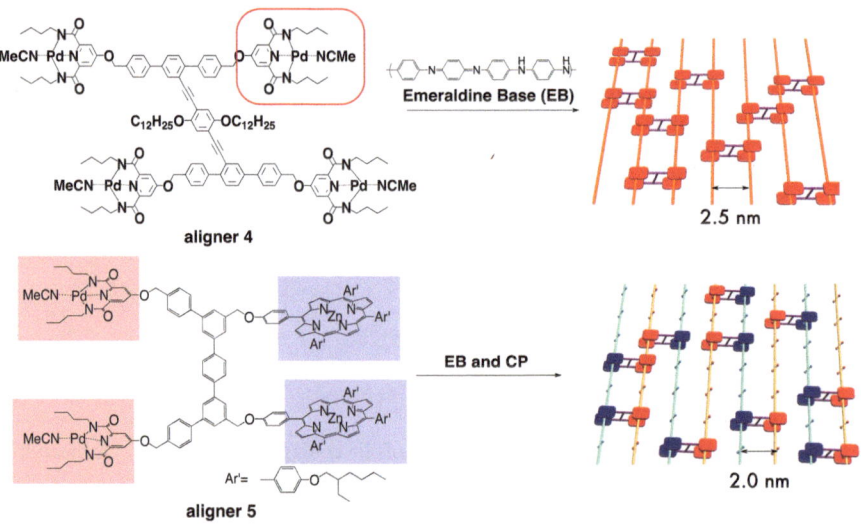

Fig. 8.4 2-D alignment of the polyanilines using aligner **4** and the construction of an alternating array of CP and polyaniline using aligner **5**

This approach essentially allowed for the construction of alternating arrays of different conjugated polymers, and was dependent on the design of the aligner molecule effectively displaying affinities towards the polymers themselves. We then proceeded to demonstrate that alternating arrays of the amino-functionalized CP and polyaniline EB were efficiently formed through the supramolecular bundling of a synthetic cross-linker **5** (Fig. 8.4) [15]. Aligner **5** possessed two different binding sites which consisted of co-facial porphyrinatozinc complexes and two palladium complexes, respectively. The former binding site was designed for recognizing CPs, whereas the latter was designed for recognizing EB through coordination bonds. These binding sites were introduced around the terphenyl rotational axis so that one binding event would exert a positive influence on the subsequent binding event. When the CP and EB were bundled and juxtaposed by aligner **5**, the distance between the two polymers was calculated to be 2.0–2.3 nm when they were in their energy-minimized states. Interestingly, the morphologies of the resultant conjugated polymer blends exhibited a well-regulated crystalline structure with a periodicity of 2.0 nm, which was clearly shown by the TEM studies. This approach can be regarded as a supramolecular 2-D polymerization of binary conjugated polymers.

8.5 Alignment of Conjugated Polymers Using the Twining Polymer as an Auxiliary

Another approach for aligning CPs involves the use of twining polymers (twimers), which can act as helical 'hosts' that effectively twine around a single CP [16]. It was envisaged that these 'hosts' would integrate chromogenic groups

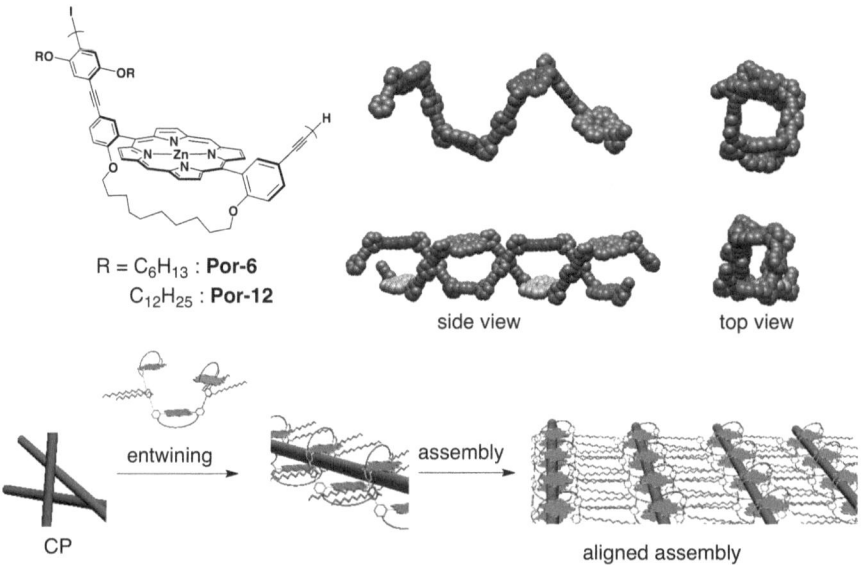

R = C₆H₁₃ : **Por-6**
C₁₂H₂₅ : **Por-12**

Fig. 8.5 Structures of **Por-6** and **Por-12** and a schematic illustration of the twimer system for the alignment of the conjugated polymer

that could mediate electron or energy transfer to or from the CP. Taking these factors into consideration, the oligomeric porphyrins **Por-12** and **Por-6** were designed as twimers (Fig. 8.5). These twimers tended to form helical structures in which a coordinative open face of the zinc porphyrin unit was always turned inwards so that the central metal atom could interact with the included CP. A decamethylene group covered the non-coordinative face. It was envisaged that a CP with the appropriate ligand groups could be become included within this helical strand through coordination with the porphyrinatozinc. The R groups in the Por oligomers were introduced to increase their solubility in organic solvents and to control the spacing between the Por/CP composites.

UV-vis absorption spectroscopic studies revealed that the ratio between the diamino units in CP and the porphyrin units in **Por-12** was 1:1, on the basis of a molar ratio plot, which suggested that only one of the two amino groups interacted with each porphyrin unit (i.e., a single-stranded **Por-12** oligomer could wrap around a single CP chain). The solid-state morphologies of [**Por-12** · CP] assemblies were examined using confocal laser scan microscopy, polarized optical microscopy (POM) and TEM. A solution-cast film of the [**Por-12** · CP] composite, which had been constructed on an indium tin oxide (ITO) glass, was subjected to POM observation, which revealed that **Por-12** and CP assemble into a highly ordered crystalline structure over the micron scale on the ITO glass. In the TEM images of the aggregate obtained from the [**Por-12** · CP] composite, several micrometer size sheets were observed with a 4.0 nm periodicity of the dark stripes, over a distance of a few hundred nanometers. Interestingly, when **Por-6** was subjected to the same conditions instead, the TEM image revealed a similar striped structure, but the distance between the stripes was shortened to only 2.7 nm. This result provided a clear indication that the distance reflected the length of the alkyl chains of the Por oligomer. In fact, if it is assumed that the partially interdigitated packing of the alkyl groups of the Por oligomers occurs in the [**Por** · CP] composites formed through the twining of Por oligomers around the CP, then it is possible to estimate the Zn^{II}–Zn^{II} distance between the adjacent composites to be 4.0 nm for the [**Por-12** · CP], and 2.7 nm for the [**Por-6** · CP] composites. These results support the view that the peripheral alkyl chains also participate in the packing among the composites.

8.6 Summary and Outlook

The concepts introduced in this chapter are complementary to the existing techniques currently available for the preparation of supramolecular and macromolecular assemblies. The introduction of interactive sites for conjugated polymers at the spatial position of the aligner or the twimer allowed for control over the dimensions, morphologies, and interpolymer spacings. The bundling of binary polymers could become a rare example of a molecular assembly technique that enables more than two polymers to assemble in a desirable fashion. The approach presented here would resolve two of the critical issues associated with conjugated polymer-based devices, including the segregation of polymers in individual chains and their controlled alignment. Furthermore, the results imply that "supramolecular bundling" (i.e., the use of molecular recognition events including coordination bonds) could play a critical role during these organization processes.

Acknowledgments M. Takeuchi would like to extend his sincerest thanks to Professor Seiji Shinkai (Sojo University), Dr. Kazunori Sugiyasu (NIMS), and the many collaborators involved in the work described in this chapter.

References

1. Skotheim TA, Elsenbaumer RL, Reynolds JR (2007) Handbook of conducting polymers, 3rd edn. CRC Press, New York
2. Günes S, Neugebauer H, Sariciftci NS (2007) Conjugated polymer-based organic solar cells. Chem Rev 107:1324–1338. doi:10.1021/cr050149z
3. Tsao HN, Müllen K (2010) Improving polymer transistor performance via morphology control. Chem Soc Rev 39:2372–2386. doi:10.1039/B918151M
4. Salleo A, Kline RJ, Delongchamp DM, Chabinyc ML (2010) Microstructural characterization and charge transport in thin films of conjugated polymers. Adv Mater 22:3812–3838. doi:10.1002/adma.200903712
5. Moons E (2002) Conjugated polymer blends: linking film morphology to performance of light emitting diodes and photodiodes. J Phys Condens Matter 14:12235–12260. doi:10.1088/0953-8984/14/47/301

6. McNeill CR, Greenham NC (2009) Conjugated-polymer blends for optoelectronics. Adv Mater 21:3840–3850. doi:10.1002/adma.200900783
7. Akagi K (2009) Helical polyacetylene: asymmetric polymerization in a chiral liquid-crystal field. Chem Rev 109:5354–5401. doi:10.1021/cr900198k
8. Kim J, Swager TM (2001) Control of conformational and interpolymer effects in conjugated polymers. Nature 411:1030–1034. doi:10.1038/35082528
9. Hamaguchi M, Yoshino K (1995) Polarized electroluminescence from rubbing-aligned poly(2,5-dinonyloxy-1,4-phenylenevinylene) films. Appl Phys Lett 67:3381–3383. doi:10.1063/1.114900
10. Kubo Y, Kitada Y, Wakabayashi R, Kishida T, Ayabe M, Kaneko K, Takeuchi M, Shinkai S (2006) A supramolecular bundling approach toward the alignment of conjugated polymers. Angew Chem Int Ed 45:1548–1553. doi:10.1002/anie.200503128
11. Takeuchi M, Ikeda M, Sugasaki A, Shinkai S (2001) Molecular design of artificial molecular and ion recognition systems with allosteric guest responses. Acc Chem Res 34:865–873. doi:10.1021/ar0000410
12. Wakabayashi R, Kubo Y, Hirata O, Takeuchi M, Shinkai S (2005) Allosteric function facilitates template assisted olefin metathesis. Chem Commun 46:5742–5744. doi:10.1039/B512805F
13. Wakabayashi R, Kubo Y, Kaneko K, Takeuchi M, Shinkai S (2006) Olefin metathesis of the aligned assemblies of conjugated polymers constructed through supramolecular bundling. J Am Chem Soc 128:8744–8745. doi:10.1021/ja063040x
14. Kaseyama T, Takebayashi S, Wakabayashi R, Shinkai S, Kaneko K, Takeuchi M (2009) Supramolecular assemblies of polyaniline through cooperative bundling by a palladium-complex-appended synthetic cross-linker. Chem Eur J 15:12627–12635. doi:10.1002/chem.200902305
15. Kaseyama T, Wakabayashi R, Shinkai S, Kaneko K, Takeuchi M (2011) Alternating arrays of different conjugated polymers utilizing a synthetic cross-linker. Chem Eur J 17:1793–1797. doi:10.1002/chem.201002675
16. Takeuchi M, Fujikoshi C, Kubo Y, Kaneko K, Shinkai S (2006) Conjugated polymers complexed with helical porphyrin oligomers create micron-sized ordered structures. Angew Chem Int Ed 45:5494–5499. doi:10.1002/anie.200601493

Chapter 9
Nanoscale Surface Science on Two-Dimensional Molecular Assembly

Soichiro Yoshimoto

Abstract The 'bottom-up strategy' is an attractive and promising approach for the construction of nanoarchitectures. This chapter will focus on advances made in the field of scanning tunneling microscopy towards the study of coordination chemistry based on surface science. A variety of different aspects of nanoscale surface science based on the 'bottom-up strategy' has been summarized in this chapter, including direct metalation and connection to surfaces, the design of nanoarchitectures consisting of porphyrin and phthalocyanine units, the direct synthesis of porphyrin oligomers on surfaces, and the identification of species that are axially coordinated to metalloporphyrin and related compounds on the surfaces.

Keywords Scanning tunneling microscopy • 2-D organic adlayers • Surface-assisted template • Axial ligand • Redox

9.1 Introduction

The construction of characteristic nanoarchitectures from organic molecules using the self-assembly technique has been the subject of considerable levels of interest [1, 2]. The use of self-assembly as a key technique for the 'bottom-up' fabrication of nanoscale functional structures in particular has received a great deal of attention ever since the concept of self assembly was established for supramolecular chemistry [1]. The two-dimensional (2-D) self-assembly of single-molecule

S. Yoshimoto (✉)
Priority Organization for Innovation and Excellence, Kumamoto University, Kumamoto, Japan
e-mail: so-yoshi@kumamoto-u.ac.jp

S. Yoshimoto
Kumamoto Institute for Photo-Electro Organics (Phoenics), Kumamoto, Japan

Y. Matsuo et al. (eds.), *Metal–Molecular Assembly for Functional Materials*,
SpringerBriefs in Molecular Science, DOI: 10.1007/978-4-431-54370-1_9,
© The Author(s) 2013

electronic devices has also attracted significant interest from the scientific community as a potential route for the fabrication of next-generation molecular electronics [3, 4]. The design of multiple-redox molecules and strategies enabling control over the peripheral intermolecular interactions as well as those between the aligned molecules and solid substrates play pivotal roles in controlling the 2-D nanostructures and patterns of the self assembled materials [5–8].

In recent years, scanning tunneling microscopy (STM) has been widely accepted as a powerful tool for understanding the structures associated with the adsorbed layers of molecules on metal surfaces at the atomic scale, under ultrahigh vacuum (UHV) [5–8] conditions, as well as in aqueous solution [6, 7]. High-resolution STM has made it possible to directly determine the packing arrangements and internal structures of adsorbed organic molecules. In situ STM allows electrode processes such as the adsorption of water-soluble inorganic and organic species to be monitored with atomic and molecular levels of resolution [7]. STM studies on the construction of 2-D metal–organic frameworks and porous networks based on coordination chemistry, represent a particularly active area of research [5–8].

Herein, we will focus on the nanoscale surface science of metal–ligand coordination based on the 'bottom-up strategy', with particular emphasis on recent STM investigations.

9.2 Direct Metalation and Connection at Surface

The construction of supramolecular assemblies using porphyrin derivatives has been investigated with the aim of potentially fabricating precisely controlled molecular wires and 2-D polymeric nanosheets. The direct synthesis of metalloporphyrins at Ag(111) has been reported by several UHV researchers [9, 10]. Following the preparation of a highly ordered adlayer of free-base H$_2$TPP on Ag(111) under UHV, the direct metalation of a porphyrin adlayer onto Ag(111) was successfully imaged via the further deposition of several other metals such as Co, Fe, and Zn onto the adlayer. The adlayers were well-characterized by UHV-STM [11], scanning tunneling spectroscopy (STS) [11], X-ray photoelectron spectroscopy (XPS) [11], and ultraviolet photoelectron spectroscopy (UPS) [12, 13]. The XPS results in particular revealed that the metal ions could be coordinated to a porphyrin ring [9, 10]. In addition, the Fe atoms could be successfully directly metalated onto the free-base phthalocyanine (H$_2$Pc) and octaethylporphyrin (H$_2$OEP) adlayers formed on both the Ag(111) and Au(111) surfaces using UHV-STM [14].

Another interesting surface reaction involves the direct synthesis of a metal-coordinated porphyrin array. From the perspective of surface science, surface-supported 2-D metal–organic frameworks (MOFs) with open-spaced adlayers have been extensively investigated under UHV on a variety of metal single crystal surfaces, such as Au, Ag, and Cu [16–18]. The 2-D structures of these MOFs, including, for example, the 2-D supramolecular coordination of not only zinc(II) tetrapyridyl porphyrin (ZnTPyP) with Au [17] but also the bi-components composed of ZnTPyP and 4′,4′′′′-(1,4-phenylene)bis(2,2′:6′,2′′-terpyridine)

(PTPy) with Fe [18], have been characterized by STM. Surface-supported metal–organic MOFs are often used as host frameworks for small organic molecules, such as fullerene [5, 16]. Organometallic 2-D nanosheets were formed on Au(111) by the reaction of tetra-cyanobenzene with Fe at elevated temperatures under UHV [19]. Thus, the choice of coordination metal makes it possible to precisely control these kinds of nanostructure.

In the field of solution-based coordination chemistry, coordination assemblies of metal ions with organic ligands, such as pyridine and carboxylic acid groups, are particularly attractive for research on surface electrochemistry because direct coordination assembly on a surface could enable the design of new nanostructures by controlling the redox potentials of the coordinated metal ions [20–22]. PTPy is often used as a molecular building block for the construction of metallosupramolecular assemblies and molecular wires in the research focused on coordination chemistry [20, 21]. The development of greater understanding of the coordination of bis(terpyridine) derivatives with metal ions on surfaces would be of significant benefit to electrochemists. In our own research, we recently focused on the self-assembly of terpyridine derivatives on Au(111) electrode surfaces. We have previously reported the direct formation of a 2-D redox-active adlayer based on PTPy and Co^{2+} on a Au(111) electrode [23]. Self-assemblies of single-component PTPy adlayers indicated the formation of the characteristic adlayer structure on Au(111), whereas PTPy and the cobalt ions simply mixed to form an uncoordinated 2-D adlayer on Au(111), as shown in Fig. 9.1. The 2-D complex-like adlayer showed electrocatalytic activity towards O_2, which was similar to the behavior observed at the cobalt(II) octaethylporphyrin-modified Au(111) electrode [24]. Surprisingly, the Co ions were stably trapped by the presence of PTPy molecules in a strong acidic solution. This method could be extended to a verity of other metal ions, such as Fe^{2+} [25].

To obtain a completely coordinated adlayer or a thin film between the ligand molecules and the Co ions, an adlayer consisting of terpyridine ligands, such as tetrapyridyl pyrazine (TPyPz) or PTPy, and cobalt ions was prepared on a Au(111) surface using a simple stepwise coordination method. The coordination abilities of these materials on Au(111) were clearly dependent upon the chemical structure of the ligand. The PTPy ligand adlayer showed that an electrochemically stable and redox-active metallosupramolecularly assembled adlayer could be directly formed on Au(111), whereas the TPyPz adlayer did not coordinate with the cobalt ions. The well-defined redox-active adlayer, which was assembled in a metallosupramolecular manner, was formed using the stepwise coordination method [26].

9.3 Surface-Assisted Molecular Adlayers

The structural patterns derived from molecular assembly processes at surfaces are mainly determined by the regulation of the relative strengths of the intermolecular and molecule–substrate interactions. An effective understanding of the relationship between the coordination ability of a metal complex and the electronic charge density of a metal single crystal surface allows us to produce a selectively patterned

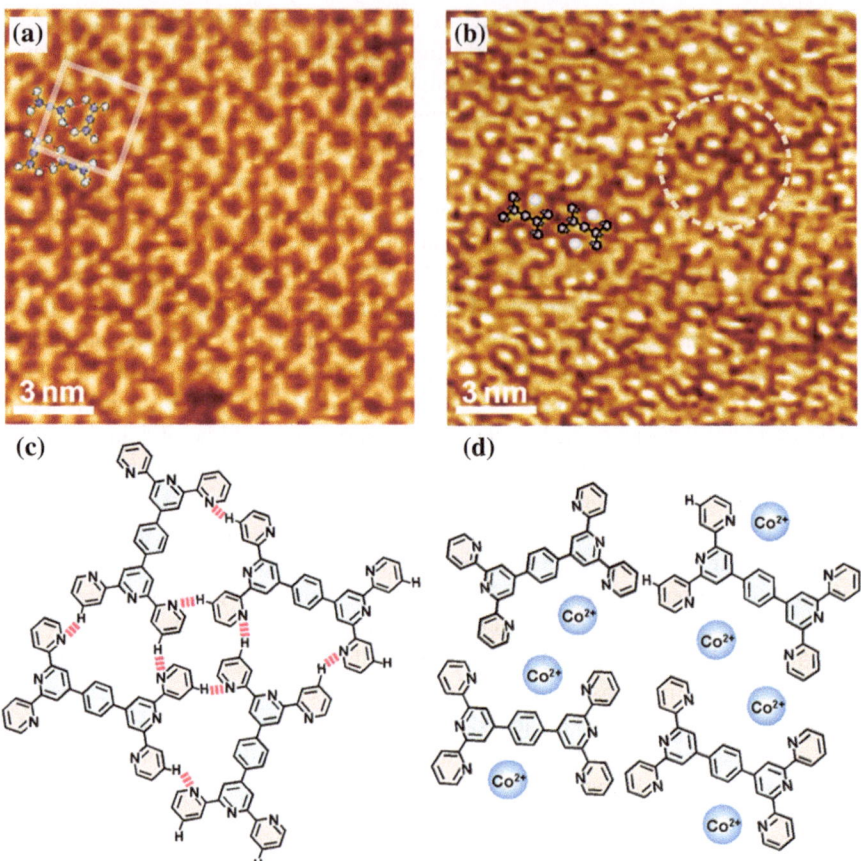

Fig. 9.1 High-resolution STM images of (**a**) a single-component PTPy and (**b**) a co-adsorbed adlayer of PTPy and Co^{2+} on Au(111) in 0.1 M $HClO_4$. (**c**) and (**d**) show the corresponding structural models for (**a**) and (**b**). Reprinted with permission from Ref. [23], Copyright (2010) The Royal Society of Chemistry

assembly on the surface. The low-index plane of a metal single crystal can act as a surface-assisted template for the molecular assembly. The reconstructed surface of a Au single crystal in particular is preferred for the selective adsorption of porphyrins and phthalocyanines. For example, we have demonstrated that the characteristic bi-molecular assemblies of copper(II) tetraphenyl porphyrin (CuTPP) and cobalt(II) phthalocyanine (CoPc) can be formed on atomic rows of (5 × 20) for the surface reconstruction of Au(100), the so-called "Au(100)–(hex)", as shown in Fig. 9.2a. On the Au(100)–(hex) surface, CuTPP and CoPc alternately aligned in a single dimension [27]. It appeared that the CuTPP molecules were located in the higher part of the reconstructed rows, whereas the CoPc molecules were located in the lower part of reconstructed rows at high tunneling currents [28]. The finding suggested that the individual CoPc and CuTPP molecules recognized their preferred adsorption sites.

(a)   **(b)**

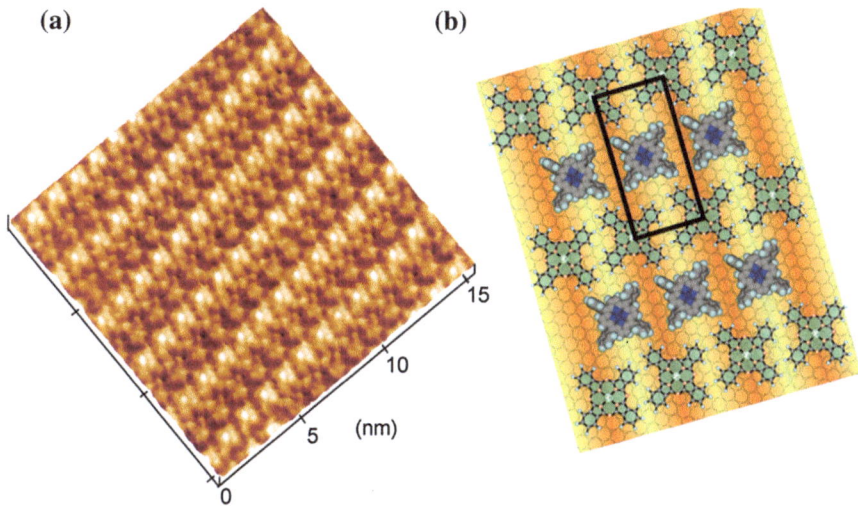

Fig. 9.2   a Height-shaded STM image and **b** The structural model of the binary adlayer consisting of CoPc and CuTPP on the reconstructed Au(100)–(hex) surface in 0.1 M HClO$_4$. Reprinted with permission from Ref. [28], Copyright (2006) American Chemical Society

Based on the STM image, a structural model of the two-component adlayer was proposed consisting of CoPc and CuTPP, as depicted in Fig. 9.2b. A rectangular lattice including two molecules has been superimposed on the image as a (5 × 10) unit cell in terms of the Au(100)–(1 × 1) structure. Each CoPc molecule is located in the valley (lowest part) of a reconstructed row, whereas each CuTPP molecule is at an elevated site of corrugation with a tilted conformation for the phenyl group. Thus, each molecule can be alternately located on the bright and dark parts of the reconstructed rows of Au(100)–(hex). This similarly one-dimensionally aligned molecular structure was also observed on Au(100)–(hex) for the bi-molecular assembly by ZnPc and ZnOEP [29]. It was subsequently found that a porphyrin derivative substituted with two bromides and iodides at positions *trans* to each other also formed a highly ordered array on the reconstructed Au(100)–(hex) surface [30]. For Au(110), the molecular adsorption of iron(II) phthalocyanine (FePc) can promote a local (1 × 5) surface reconstruction of Au(110), driving the assembly of molecular chains in the direction of the atomic lattice, which is the so-called [1$\bar{1}$0] direction. Density functional theory (DFT) calculations supported the energetic origins of the molecule-driven substrate reconstruction [31].

In addition, the metal single crystal plane of (110) was found to be available for the formation of a supramolecularly assembled porphyrin and C$_{60}$ adlayer [32], as well as organometallic 1-D oligomers [33]. The direct coupling reaction of porphyrin derivatives, in particular, took place on the Cu(110) aligning the [001] direction under elevated temperatures in the range of 560–650 K [33]. Under these high temperature conditions, the coupling reaction between the porphyrin derivatives was readily accelerated with mobile Cu atoms on the Cu(110) surface.

The metal single crystal surface, especially the (110) plane, could therefore serve as a molecular template. The technique used for the connection of molecules at the surface is important for the preparation of covalently bound 2-D arrays.

9.4 Identification of Coordination Axial Ligand

Metalloporphyrins have been recognized as an important material in a number of particularly important technologies, such as gas-sensing, catalytic, photo-electronic and fuel cell systems. It is also known that porphyrins and related derivatives constitute active centers, which are otherwise known as 'hemes', in metalloproteins such as myoglobin and hemoglobin. A number of different 'picket–fence' type porphyrins ($MT_{piv}PP$) in particular were synthesized by Collman's group [34], and they were found to serve as model systems for understanding the process of dioxygen storage. The O_2 binding affinities of $FeT_{piv}PP$ and $CoT_{piv}PP$ have been demonstrated to be much greater than that of native myoglobin. The binding affinity of O_2 is controlled by the coordination of this axial ligand to the vacant sixth coordination site [35]. It is necessary to understand the role and the electrochemical properties of these materials in relation to certain phenomena, such as the storage of dioxygen. Several interesting phenomena were observed for the 'picket–fence' porphyrins adlayers on the Au single crystal surfaces [36]. The state of the molecular oxygen trapped in the cavity of $CoT_{piv}PP$ was distinctly observed in STM images as a bright spot in the nanobelt array formed on the reconstructed Au(100)–(hex) surface, but not on the Au(111) surface. These results suggested that the arrangement of underlying Au atoms plays an important role on the sixth ligand coordination sites assisting the formation of the O_2-adducted $CoT_{piv}PP$. Figure 9.3a reveals that the bright spots observed in the STM images can be attributed to the O_2-adducted $CoT_{piv}PP$ molecules. The potential manipulation allowed for the O_2 trapped inside the $CoT_{piv}PP$ cavities to be released onto the nanobelt array, as shown in Fig. 9.3c–e.

It is well known that the conformational flexibility of the porphyrin framework allows for the relevant functional properties to be effectively regulated. Several interesting investigations have been performed on ligand coordination [37–39]. To develop a precise understanding of the interplay mechanism between the structural distortions of the porphyrin framework and the ligation of small adducts, the carbon monoxide (CO) axial ligand was clearly identified in the saddle-type CoTPP and FeTPP arrays formed on Ag(111) under UHV [39].

In contrast to the porphyrin and phthalocyanine complexes, mixed valence-state complexes, such as those exemplified by the 1-D halogen-bridged mixed-valence compounds, are of particular interest to a variety of different disciplines [40–42], because they exhibit novel electric, magnetic and optical properties on account of their unique oxidation and spin states. String-like 1-D coordinated polymer chains made from halogen-bridged diruthenium complexes were observed on a highly orientated pyrolytic graphite surface using STM and atomic force microscopy [43, 44]. The significance of such mixed valence complexes can be found in their application

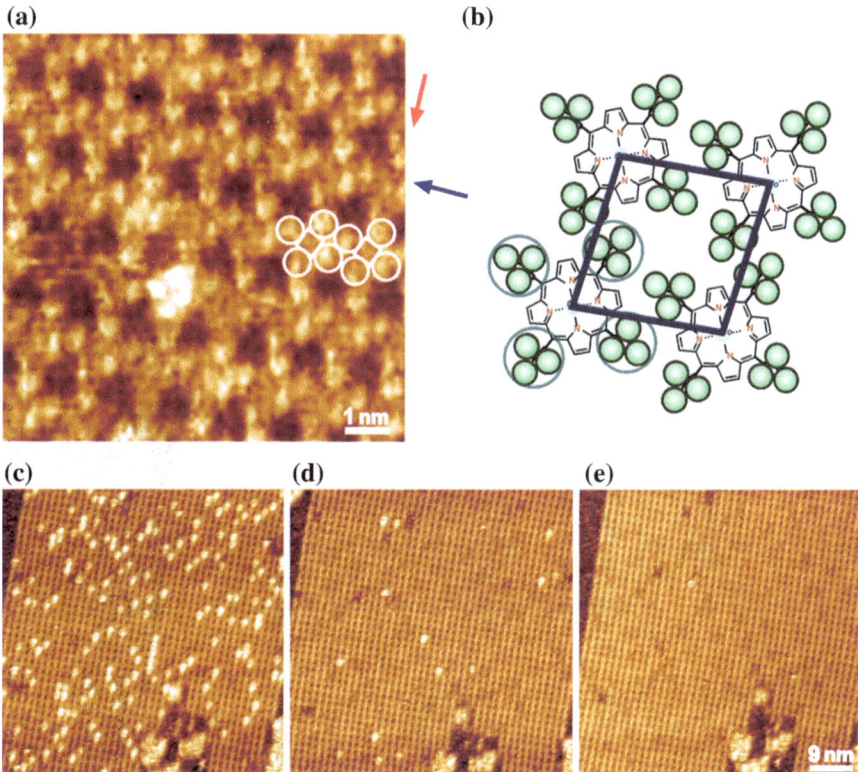

Fig. 9.3 **a** High-resolution STM image and **b** The corresponding proposed model of a highly ordered array of CoT$_{piv}$PP on Au(111) in HClO$_4$. (**c–e**) Potential-dependent STM images of the highly ordered CoT$_{piv}$PP nanobelt array formed on Au(100)–(hex) at the same location. The images were obtained at (**c**) 0.80 V, (**d**) 0.73 V, and (**e**) 0.68 V versus RHE in 0.1 M HClO$_4$. Reprinted with permission from Ref. [36], Copyright (2007) American Chemical Society

to molecular computing system, i.e., the quantum-dot cellular automata (QCA) cell [45–47]. The QCA cell enables the writing and reading of molecular memory on the nanoscale size level [46]. In contrast, paddle-wheel type mixed valence complexes with two perpendicularly stacked ruthenium centers are considered to be ideal systems for the development of high-density molecular memories, because they show multiple redox states (i.e., Ru^{II}/Ru^{II} and Ru^{II}/Ru^{III}-X, X = halogen ion) which are expected to allow for the writing and reading of electronic signals over a minimum molecular cross-sectional area.

The 2-D molecular assemblies of chloride-coordinated mixed-valence diruthenium complexes, each possessing phenyl, naphthyl, or anthracenyl moieties, were examined on Au(111) at the electrochemical interface (see Fig. 9.4a) [48]. The results revealed the presence of a characteristic redox wave of complex **1** in the potential region between 0.75 and 0.10 V. The amount of transferred electronic charge was estimated via the integration of the peak area, and the average

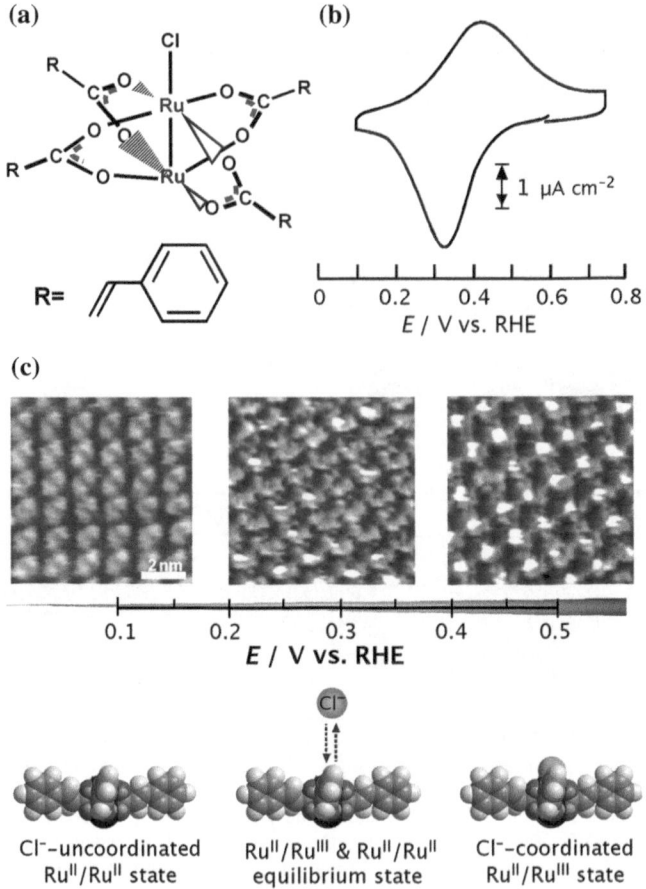

Fig. 9.4 **a** Chemical structure of the paddle-wheel diruthenium complex **1**. **b** Cyclic voltammogram of **1**-adsorbed onto a Au(111) electrode in 0.1 M HClO$_4$ recorded at a scan rate of 20 mV s^{-1}. **c** Potential-dependent STM images and their proposed model of the **1** adlayer on Au(111) observed in 0.1 M HClO$_4$. Reprinted with permission from Ref. [48], Copyright (2012) American Chemical Society

value was determined to be 1.12×10^{-5} C cm^{-2}. This value leads to a surface excess of 1.16×10^{-10} mol cm^{-2}, assuming that a single-electron reaction occurs on the **1**-modified Au(111) electrode. Figure 9.4c provides an overview of the potential-dependent in situ STM images of the paddle-wheel diruthenium complex on Au(111). The Ru^{II}/Ru^{III} and Ru^{II}/Ru^{II} redox states can coexist in the potential region. At potentials where the Ru^{II}/Ru^{III} and Ru^{II}/Ru^{II} redox states were in equilibrium, two distinct redox states were clearly identified at the single-molecular level. A Cl$^-$ exchange process took place in the 2-D ordered adlayer of **1** during the scan, suggesting that Cl$^-$ has a potential-controlled weak interaction with the diruthenium complex, as depicted in the models shown in Fig. 9.4c. This chloride ion exchange would be accomplished via an equilibration between the oxidized and reduced forms of complex **1**. The coexistence of these two redox states results from a rapid internal electron exchange between the Ru^{II}/Ru^{III} and Ru^{II}/Ru^{II} complexes in the highly ordered adlayer of complex **1**.

Most recently, a similar phenomenon was observed in a highly ordered array of chloride-coordinated iron tetraphenylporphyrin (FeClTPP) on Au(111) using UHV-STM at low-temperature [49]. By manipulating the STM tip, Cl could be selectively removed from the central Fe ion of a FeClTPP molecule without disrupting the surrounding molecular pattern. This Cl abstraction from FeClTPP was triggered by the removal of an electron from the highest occupied molecular orbital. DFT calculations suggested that the reaction involved a change in the oxidation state of the Fe ion. This reaction mechanism is principally consistent with the redox control of the central metal ion under electrochemical conditions.

9.5 Summary and Outlook

In this chapter, we have summarized the recent advances in the nanoscale surface science of coordination chemistry based on STM studies. Metal–ligand coordination not only plays an important role in the regulation of 2-D porous networks but also plays a role in the regulation of redox potential and electrocatalytic activity. Developing a greater understanding of the influences of conformational change, direct connection, and ligand exchange at the nanoscale will enable us to create new characteristic nanoarchitectures and devices. It is envisaged that solution-based coordination chemistry will continue to make further contributions to the construction of functional nanoarchitectures.

Acknowledgments   The author would like to express his sincerest thanks to Prof. Kingo Itaya (Tohoku University, Japan) and the many collaborators involved in the work described above.

References

1. Lehn JM (1995) Supramolecular Chemistry. Wiley-VCH, Weinheim
2. Li G, Fudickar W, Skupin M, Klyszcz A, Draeger C, Lauer M, Fuhrhop JH (2002) Rigid lipid membranes and nanometer clefts: motifs for the creation of molecular landscapes. Angew Chem Int Ed 41:1828–1852. doi:10.1002/1521-3773(20020603)41:11<1828:AID-ANIE1828>3.0.CO;2-#
3. Jortner J, Ratner MA (1997) Molecular Electronics. Blackwell Science, Oxford
4. Joachim C, Gimzewski JK, Aviram A (2000) Electronics using hybrid-molecular and mono-molecular devices. Nature 408:541–548. doi:10.1038/35046000
5. Barth JV, Costantini G, Kern K (2005) Engineering atomic and molecular nanostructures at surfaces. Nature 437:671–679. doi:10.1038/nature04166

6. Kudernac T, Lei S, Elemans JAAW, De Feyter S (2009) Two-dimensional supramolecular self-assembly: nanoporous networks on surfaces. Chem Soc Rev 38:402–421. doi:10.1039/b708902n

7. Yoshimoto S, Kobayashi N (2010) Supramolecular nanostructures of phthalocyanines and porphyrins at surfaces based on the "bottom-up assembly". Struc Bond 135:137–168. doi:10.1007/978-3-642-04752-75

8. Bartels L (2010) Tailoring molecular layers at metal surfaces. Nature Chem 2:87–95. doi:10.1038/nchem.517

9. Gottfried JM, Flechtner K, Kretschmann A, Lukasczyk T, Steinrück HP (2006) Direct synthesis of a metalloporphyrin complex on a surface. J Am Chem Soc 128:5644–5645. doi:10.1021/ja0610333

10. Kretschmann A, Walz MM, Flechtner K, Steinrück HP, Gottfried JM (2007) Tetraphenylporphyrin picks up zinc atoms from a silver surface. Chem Commun 568–570. doi:10.1039/b614427f

11. Buchner F, Flechtner K, Bai Y, Zillner E, Kellner I, Steinrück HP, Marbach H, Gottfried JM (2008) Coordination of iron atoms by tetraphenylporphyrin monolayers and multilayers on Ag(111) and formation of iron-tetraphenylporphyrin. J Phys Chem C 112:15458–15465. doi:10.1021/jp8052955

12. Lukasczyk T, Flechtner K, Merte LR, Jux N, Maier F, Gottfried JM, Steinrück HP (2007) Interaction of Cobalt(II) tetraarylporphyrins with a Ag(111) surface studied with photoelectron spectroscopy. J Phys Chem C 111:3090–3098. doi:10.1021/jp0652345

13. Comanici K, Buchner F, Flechtner K, Lukasczyk T, Gottfried JM, Steinrück HP, Marbach H (2008) Understanding the contrast mechanism in scanning tunneling microscopy (STM) images of an intermixed tetraphenylporphyrin layer on Ag(111). Langmuir 24:1897–1901. doi:10.1021/la703263e

14. Bai Y, Buchner F, Wendahl MT, Kellner I, Bayer A, Steinrück HP, Marbach H, Gottfried JM (2008) Direct metalation of a phthalocyanine monolayer on Ag(111) with coadsorbed iron atoms. J Phys Chem C 112:6087–6092. doi:10.1021/jp711122w

15. Bai Y, Sekita M, Schmid M, Bischof T, Steinrück HP, Gottfried JM (2010) Interfacial coordination interactions studied on cobalt octaethylporphyrin and cobalt tetraphenylporphyrin monolayers on Au(111). Phys Chem Chem Phys 12:4336–4344. doi:10.1039/b924974p

16. Stepanow S, Lin N, Barth JV (2008) Modular assembly of low-dimensional coordination architectures on metal surfaces. J Phys Condens Matter 20:184002 (1–15). doi:10.1088/0953-8984/20/18/184002

17. Shi Z, Lin N (2009) Porphyrin-based two-dimensional coordination Kagome lattice self-assembled on a Au(111) surface. J Am Chem Soc 131:5376–5377. doi:10.1021/ja900499b

18. Shi Z, Lin N (2010) Structural and chemical control in assembly of multicomponent metal-organic coordination networks on a surface. J Am Chem Soc 132:10756–10761. doi:10.1021/ja1018578

19. Abel M, Clair S, Ourdjini O, Mossoyan M, Porte L (2011) Single layer of polymeric Fe-phthalocyanine: an organometallic sheet on metal and thin insulating film. J Am Chem Soc 133:1203–1205. doi:10.1021/ja108628r

20. Díaz DJ, Storrier GD, Bernhard S, Takada K, Abruña HD (1999) Ordered arrays generated via metal-initiated self-assembly of terpyridine containing dendrimers and bridging ligands. Langmuir 15:7351–7354. doi:10.1021/la990513n

21. Nishihara H, Kanaizuka K, Nishimori Y, Yamanoi Y (2007) Construction of redox- and photo-functional molecular systems on electrode surface for application to molecular devices. Coord Chem Rev 251:2674–2687. doi:10.1016/j.ccr.2007.04.002

22. Han F, Higuchi M, Kurth D (2008) Metallosupramolecular polyelectrolytes self-assembled from various pyridine ring-substituted bisterpyridines and metal ions: photophysical, electrochemical, and electrochromic properties. J Am Chem Soc 130:2073–2081. doi:10.1021/ja710380a

23. Yoshimoto S, Ono Y, Nishiyama K, Taniguchi I (2010) Direct formation of a 2D redox-active adlayer based on a bisterpyridine derivative and Co^{2+} on a Au(111) electrode. Phys Chem Chem Phys 12:14442–14444. doi:10.1039/c0cp00981d

24. Yoshimoto S, Inukai J, Tada A, Abe T, Morimoto T, Osuka A, Furuta H, Itaya K (2004) Adlayer structure of and electrochemical O_2 reduction on cobalt porphine-modified and

cobalt octaethylporphyrin-modified Au(111) in $HClO_4$. J Phys Chem B 108:1948–1954. doi:10.1021/jp0366421

25. Nishiyama K, Ono Y, Taniguchi I, Yoshimoto S (2012) EC-STM investigation of electrochemically active 2D adlayer consisting of metal ions and a bisterpyridine derivative. Chem Lett 41:1311–1313. doi:10.1246/cl.2012.1311

26. Yoshimoto S, Nishiyama K (2013) One-pot formation of a metallosupramolecularly assembled and redox-active adlayer at the solid–liquid interface. J Inorg Organomet Polym Mater 23:233–238. doi:10.1007/s10904-012-9743-3

27. Suto K, Yoshimoto S, Itaya K (2003) Two-dimensional self-organization of phthalocyanine and porphyrin: dependence on the crystallographic orientation of Au. J Am Chem Soc 125:14976–14977. doi:10.1021/ja038857u

28. Suto K, Yoshimoto S, Itaya K (2006) Electrochemical control of the structure of two-dimensional supramolecular organization consisting of phthalocyanine and porphyrin on a gold single crystal surface. Langmuir 22:10766–10776. doi:10.1021/la061257z

29. Yoshimoto S, Honda Y, Ito O, Itaya K (2008) Supramolecular pattern of fullerene on 2D bimolecular 'chessboard' consisting of bottom-up assembly of porphyrin and phthalocyanine molecules. J Am Chem Soc 130:1085–1092. doi:10.1021/ja077407p

30. Lafferentz L, Eberhardt V, Dri C, Africh C, Comelli G, Esch F, Hecht S, Grill L (2012) Controlling on-surface polymerization by hierarchical and substrate-directed growth. Nature Chem 4:215–220. doi:10.1038/nchem.1242

31. Fortuna S, Gargiani P, Betti MG, Mariani C, Calzolari A, Modesti S, Fabris S (2012) Molecule-driven substrate reconstruction in the two-dimensional self-organization of Fe-phthalocyanines on Au(110). J Phys Chem C 116:6251–6258. doi:10.1021/jp211036m

32. Sedona F, Di Marino M, Sambi M, Carofiglio T, Lubian E, Casarin M, Tondello E (2010) Fullerene/porphyrin multicomponent nanostructures on Ag(110): from supramolecular self-assembly to extended copolymers. ACS Nano 4:5147–5154. doi:10.1021/nn101161a

33. Haq S, Hanke F, Dyer MS, Persson M, Iavicoli P, Amabilino DB, Raval R (2011) Clean coupling of unfunctionalized porphyrins at surfaces to give highly oriented organometallic oligomers. J Am Chem Soc 133:12031–12039. doi:10.1021/ja201389u

34. Collman JP, Gagne RR, Halbert TR, Marchon JC, Reed CA (1973) Reversible oxygen adduct formation in ferrous complexes derived from a picket fence porphyrin. Model for oxymyoglobin. J Am Chem Soc 95:7868–7870. doi:10.1021/ja00804a054

35. Zou S, Clegg RS, Anson FC (2002) Attachment of cobalt "picket fence" porphyrin to the surface of gold electrodes coated with 1-(10-mercaptodecyl)imidazole. Langmuir 18:3241–3246. doi:10.1021/la011444r

36. Yoshimoto S, Sato K, Sugawara S, Chen Y, Ito O, Sawaguchi T, Niwa O, Itaya K (2007) Formation of supramolecular nanobelt arrays consisting of cobalt(II) "picket-fence" porphyrin on Au surfaces. Langmuir 23:809–816. doi:10.1021/la0617331

37. Flechtner K, Kretschmann A, Steinrück HP, Gottfried JM (2007) NO-induced reversible switching of the electronic interaction between a porphyrin-coordinated cobalt ion and a silver surface. J Am Chem Soc 129:12110–12111. doi:10.1021/ja0756725

38. Buchner F, Seufert K, Auwärter W, Heim D, Barth JV, Flechtner K, Gottfried JM, Steinrück HP, Marbach H (2009) NO-induced reorganization of porphyrin arrays. ACS Nano 3:1789–1794. doi:10.1021/nn900399u

39. Seufert K, Bocquet ML, Auwärter W, Weber-Bargioni A, Reichert J, Lorente N, Barth JV (2011) Cis-dicarbonyl binding at cobalt and iron porphyrins with saddle-shape conformation. Nature Chem 3:114–119. doi:10.1038/nchem.956

40. Kitagawa H, Mitani T (1999) Valence transition with charge ordering in a conductive MMX-chain complex. Coord Chem Rev 190–192:1169–1184. doi:10.1016/S0010-8545(99)00174-5

41. Mikuriya M, Yoshioka D, Honda M (2006) Magnetic interactions in one-, two-, and three-dimensional assemblies of dinuclear ruthenium carboxylates. Coord Chem Rev 250:2194–2211. doi:10.1016/j.ccr.2006.01.011

42. Kuwahara R, Fujikawa S, Kuroiwa K, Kimizuka N (2012) Controlled polymerization and self-assembly of halogen-bridged diruthenium complexes in organic media and their dielectrophoretic alignment. J Am Chem Soc 134:1192–1199. doi:10.1021/ja208958t

43. Olea D, González-Prieto R, Priego JL, Barral MC, de Pablo PJ, Torres MR, Gómez-Herrero J, Jimánez-Aparicio R, Zamora F (2007) MMX polymer chains on surfaces. Chem Commun 1591–1593. doi:10.1039/b613836e

44. Welte L, González-Prieto R, Olea D, Torres MR, Priego JL, Jiménez-Aparicio R, Gómez-Herrero J, Zamora F (2008) Time-dependence structures of coordination network wires in solution. ACS Nano 2:2051–2056. doi:10.1021/nn800439v

45. Braun-Sand SB, Wiest O (2003) Theoretical studies of mixed-valence transition metal complexes for molecular computing. J Phys Chem A 107:285–291. doi:10.1021/jp0265945

46. Li ZH, Fehlner TP (2003) Molecular QCA cells. 2. Characterization of an unsymmetrical dinuclear mixed-valence complex bound to a Au surface by an organic linker. Inorg Chem 42:5715–5721. doi:10.1021/ic026255q

47. Jiao J, Long GJ, Rebbouh L, Grandjean F, Beatty AM, Fehlner TP (2005) Properties of a mixed-valence $(Fe^{II})_2(Fe^{III})_2$ square cell for utilization in the quantum cellular automata paradigm for molecular electronics. J Am Chem Soc 127:17819–17831. doi:10.1021/ja0550935

48. Yoshimoto S, Sakata K, Kuwahara R, Kuroiwa K, Kimizuka N, Kunitake M (2012) Electrochemically controlled 2D assembly of paddle-wheel diruthenium complexes on the Au(111) surface and identification of their redox states. J Phys Chem C 116:17729–17733. doi:10.1021/jp305951d

49. Gopakumar TG, Tang H, Morillo J, Berndt R (2012) Transfer of Cl ligands between adsorbed iron tetraphenylporphyrin molecules. J Am Chem Soc 134:11844–11847. doi:10.1021/ja302589z